RISK-BASED DECISION MAKING IN WATER RESOURCES VIII

PROCEEDINGS OF THE EIGHTH CONFERENCE

October 12–17, 1997
Santa Barbara, California

SPONSORED BY
Engineering Foundation

CO-SPONSORED BY
Universities Council on Water Resources
Task Committee on Risk Analysis and Management of the
Committee on Water Resources Planning of the
ASCE Water Resources Planning and Management Division

SUPPORTED BY
National Science Foundation
U.S. Army Corps of Engineers, Institute for Water Resources

APPROVED FOR PUBLICATION BY
Water Resources Planning and Management Division of the
American Society of Civil Engineers

EDITED BY
Yacov Y. Haimes, David A. Moser, and Eugene Z. Stakhiv

TECHNICAL EDITOR
Grace I. Zisk

ASCE American Society
of Civil Engineers
1801 ALEXANDER BELL DRIVE
RESTON, VIRGINIA 20191–4400

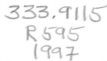

Abstract: This volume, *Risk-Based Decision Making in Water Resources VIII*, comprises edited papers from the proceedings of the Eighth Engineering Foundation Conference held in Santa Barbara, California, on October 12-17, 1997. Participants from government, industry, and universities shared their knowledge and experience in the rapidly growing field of risk-based decision making in water resources. Although many of the topics and lingering issues were addressed in at least one or more of the previous seven conferences, a sense of maturity prevailed during the five days of intensive sessions. Legislative initiatives and risk assessment, climate change, El Niño, maintenance and rehabilitation, reliability of physical infrastructure, and the risk of extreme events were among the main topics that dominated the discussion. Intensive brainstorming sessions revealed what many participants have known earlier: as we succeed in developing theory and methodology in risk assessment and management and apply them to water resources problems, much more remains to be addressed in future conferences and research.

ISBN: 0-7844-0347-3

PREFACE

These proceedings of the eighth Engineering Foundation Conference on Risk-Based Decision Making in Water Resources illustrate that our original objectives and goals, as well as the issues raised, remain relevant, important, and timely.

In the preface to the third conference, held in November 1987, we asked: "What has changed during the past decade to warrant yet another conference?" To answer this question, we stated that the conference's goals and objectives were to:

1) familiarize the participants with the state-of-the-art in risk/benefit analysis,

2) explore the feasibility of using risk/benefit analysis in water resources planning and management,

3) provide a medium conducive to the exchange of information on the conference theme among educators, analysts, managers, and policy makers, and

4) identify and articulate future desired actions designed to alleviate some of the present problems we face in risk/benefit analysis and risk assessment in general.

What had changed in 1987 was that in this still-new professional niche, we had matured. Those of us involved in risk-based decision making experienced the same evolutionary process that systems analysts and systems engineers went through earlier. In fact, in 1987 there were many who simply saw risk analysis as a specialized extension of the body of knowledge and evaluation perspectives that had come to be associated with systems analysis.

Today, as more than a decade ago, there is a strong public awareness of the subject of risk: environmental risks, technological and natural risks, and human health and safety risks. The professional community is responding much more forcefully and knowledgeably as well, and in many instances, leading what has ultimately come to be a political debate. We are more critical of the tools that we have developed, because we recognize their ultimate importance and usefulness in the resolution of critical societal problems. We are more willing to accept the premise that a truly effective risk analysis study must, in most cases, be cross-disciplinary, relying on social and behavioral scientists, engineers, regulators, and lawyers.

In 1987, we were convinced that risk assessment and management must be an integral part of the decision-making process, rather than a gratuitous add-on technical analysis. Some of us were becoming more and more convinced of the grave limitations of the traditional and commonly-used expected value concept, and were complementing and supplementing this concept with conditional expectations, where decisions about extreme and catastrophic events are not averaged out with more commonly occurring high-frequency/low-consequence events. These are some of the trends that distinguished the conference of 1987 from those of 1980 and 1985, and continue today.

During the fourth conference in October 1989, we searched for the driving forces leading to the growing popularity of risk-based decision making. We attributed the increasing popularity and prominence of risk analysis to two basic factors: society and technology.

At the opening of the fifth conference in November 1991, we quoted a chief executive officer who told his board of directors: "Our role is to manage change; if we can't manage change, we must change management." Noting that "the field of risk analysis is changing fast and we must not be left behind," we stated:

• We meet in these conferences to exchange information and knowledge on the changes that are taking place in the field.
• We must facilitate the process of listening to each other in these meetings by getting to know each other on a personal basis.
• We must create an environment that is conducive to dialogue and communications: fewer lectures and more discussions.
• We must open new lines of communication.
• We must be able to challenge ourselves and let others challenge our old assumptions.
• We must free ourselves of prejudices.
• We must be ready to reexamine our biases – professional , personal, and other biases.
• We must be able to learn from each other, to discover what new theories and methodologies have been developed and where have they been applied – either successfully or with less success.
• We should be ready to accept the premise that risk management must be an integral part of a total systems management, and adopt a holistic philosophy.

The sixth conference, held in 1993, continued to reinforce the Socratic culture that has evolved over the last two-and-a-half decades in these meetings. Although some of the papers covered topics presented previously, the discussions were more substantive and in greater depth. Methodologies were more closely related to theory, and at the same time the relevance of their applications to emerging natural and man-made hazards became stronger and more convincing. Such topics as uncertainties in data, models, and forecasts and their influences on risk analysis have, in some sense, an eternal life of their own; yet the level of discussion epitomized the growth and maturity in the field.

The seventh conference, in October 1995, augmented the technical discussion with policy issues and the implications of recent legislative initiatives in risk assessment. It attempted to address the connectedness among such emerging trends and ideas as the management of our environment, physical infrastructure, response to possible climate change, the desire to embrace the concept of sustainable development in its broader sense, and the explosion of communications opportunities and their impact on informed decision making.

The eighth Engineering Foundation Conference on Risk-Based Decision Making, held October 12-17, 1997 in Santa Barbara, California, continued exploring these themes, and also focused on the climatic effects of El Niño.

We hope that the interest in these Engineering Foundation Conferences remains as high as it has been in the past and we look forward to the ninth conference in 1999.

Several organizations and individuals were instrumental in making this conference possible. Thanks to Dr. Eleonora Sabadell, Program Manager of the National Science Foundation Research Program "Natural and Man-Made Hazards," for her essential support of this conference and for her encouragement. Kyle E. Schilling, Director of the US Army Institute for Water Resources, was equally generous in providing financial support for the conference, as was Dr. Charles Freiman of the Engineering Foundation.

We thank, also, the ASCE Task Committee on Risk-Based Decision Making and the Universities Council on Water Resources for again co-sponsoring this conference and enabling a large number of prominent speakers and participants to present and exchange ideas. Ultimately, the value of such a conference lies in the influence that these ideas and their presenters have on the vital issues and frequently extraordinary events occurring in our rapidly changing world.

All papers have been reviewed, edited, and accepted for publication in these proceedings by the editors. The papers are eligible for discussion in the *Journal of Water Resources Planning and Management* and also eligible for ASCE awards.

Finally, we acknowledge the invaluable editorial work provided by Grace I. Zisk, the administrative assistance provided by Leslie Yowell, Manager, Center for Risk Management of Engineering Systems, University of Virginia, the staff of the Engineering Foundation, and Charlotte McNaughton, Manager of Book Production of the ASCE, for her hard work in bringing these proceedings to their final printed form.

Yacov Y. Haimes
Charlottesville, Virginia

David A. Moser and Eugene Z. Stakhiv
Fort Belvoir, Virginia

Contents

is successful, then the worst consequence is a shutdown of service. If late detection is unsuccessful but emergency response is adequate, the worst consequence is human ingestion. If all of the mitigating systems are unsuccessful, then the worst consequence is human illness or death.

Such an event tree (Ang and Tang 1984; Kumamoto and Henley 1996) is useful to enumerate the possible repercussions of a given initiating event. Given the chemical tampering, there are six possibilities of worst consequences according to the event tree: warning, missed detection, system damage, service loss, human ingestion, and illness or death.

Risk assessment proceeds with estimating the likelihoods of occurrence of the various paths of the tree. Each branching in the event tree has associated probabilities of success (upper path) and failure (lower path). The sum of the probabilities of all paths is one. The probabilities of taking the various paths of the tree may be estimated through detailed modeling of the behavior of the system.

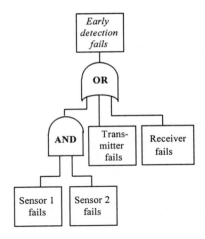

Figure 4. Fault Tree for the Failure of Early Detection of a
Chemical Contaminant Introduced at a Tap

For instance, Figure 4 indicates that early detection fails if both of two sensors fail or either transmission of the message of detection or its reception by an authority fails. For example, the two sensors could be an idealization of a water customer's abilities to taste and smell the contaminant. The fault tree (Ang and Tang 1984; Kumamoto and Henley 1996) of Figure 4 depicts the potential failures of the various components associated with early detection of a contaminant. The logic of the fault tree describes how higher-level system failures come about as the result of lower-level component failures. Knowledge of the fault tree together with the probabilities of the lowest-level failures is used to estimate the probabilities of higher-level failures. For this case,

Prob[Early detection fails]
= Prob[(Sensor 1 fails $\cap$ Sensor 2 fails) $\cup$ Transmitter fails $\cup$ Receiver fails]

where $\cap$ and $\cup$ indicate respectively the intersection and union of events. The higher-level failure of interest in the Figure 4 fault tree is the first branching of the Figure 3 event tree. In addition to fault trees, Markov models, resistance-loading models, Monte Carlo simulations, or other probabilistic models may be useful to estimate the probabilities of the various paths of an event tree.

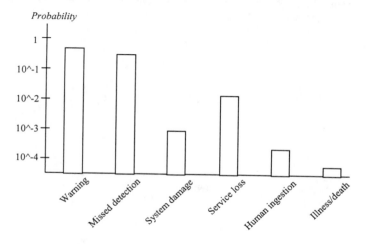

Figure 5. Probability Density Function of the Worst Consequences of Chemical Tampering at a Tap

Figure 5 gives a probability density function of the possible consequences of chemical tampering. Note that in general the probabilities of the various consequences can vary by many orders of magnitude. The probability density function in Figure 5 is a characterization of the risk conditional on the occurrence of chemical tampering at a tap of the water system.

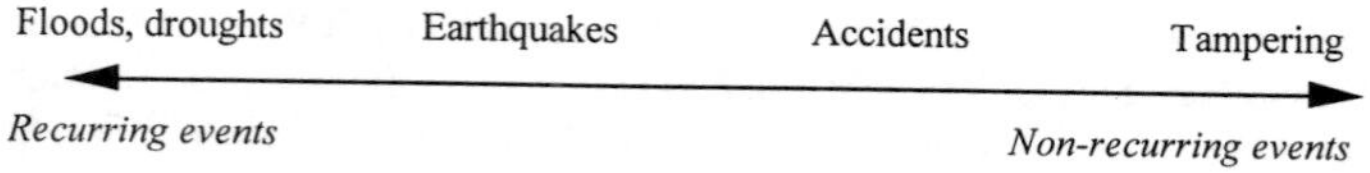

Floods, droughts   Earthquakes   Accidents   Tampering

←———————————————————————————————————→

Recurring events   *Non-recurring events*

Figure 6. Variety of Natural and Man-Made Hazards to Water Systems

Conclusions

The likelihood of chemical tampering occurring may not be known. Moreover, the tampering is the act of a willful adversary. Such acts of the adversary may be influenced by the actions of the system operator or even by an analyst estimating the likelihood of tampering. Figure 6 depicts a range of natural and man-made hazards to which water systems are subjected. Floods and droughts, to the left of the spectrum, are often considered to be recurring events and are readily addressed in terms of their frequencies of occurrence (e.g., mean occurrences per year, mean waiting time between events). Time-series data of recurring floods and droughts are often plentiful. On the other hand, earthquakes and accidents may be unique, non-recurring, and thus difficult to model from a perspective of frequency. Furthermore, tampering, which lies to the right of the range, may be even more inaccessible from the concept of a frequency: (1) there is a willful actor behind possible events, and (2) there is a perceived uniqueness to every such past and potential event.

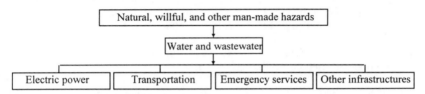

Figure 7. Water Systems' Influence on Other Infrastructures

Furthermore, Figure 7 illustrates that the failures of water systems threaten the well-being of other infrastructure systems. An effort is needed to characterize the linkages of water to other infrastructures with respect to tampering. Haimes et al. (1990) address the hierarchical multiobjective analysis of large-scale systems. Duckstein and Parent (1994) and Ganoulis et al. (1991) address the management of risk in water resources from a perspective of systems engineering.

Acknowledgments

The author wishes to thank Professor Yacov Y. Haimes and Dr. Nicholas C. Matalas of the University of Virginia, and Captains Barry C. Ezell and Robert L. Watson of the US Army and the Department of Systems Engineering, University of Virginia, for their contributions of ideas for this paper.

Support for this study was provided in part by the National Science Foundation under grant CMS-9526056, "Risk-Based Sustainable Policy for Distributed Flood Protection."

8   RISK-BASED DECISION MAKING IN WATER RESOURCES VIII

References

Ang, A-H.S., and W. Tang. 1984. *Probability Concepts in Engineering Planning and Design: Vol. 2*. New York: John Wiley and Sons.

Clahane, D. 1997. Dealing with a radioactive contamination of the water supply. *Water Management International* (the annual review of the water and wastewater industry). London, England: Sterling Publications, Ltd. 85-87.

Duckstein, L., and E. Parent. 1994. Systems engineering of natural resources under changing physical conditions: a framework for reliability and risk. In L. Duckstein and E. Parent (Eds.), *Engineering Risk in Natural Resources Management* (NATO ASI Series). Boston, MA: Kluwer Academic Publishers.

Ganoulis, J., L. Duckstein, and I. Bogardi. 1991. Risk analysis of water quantity and quality problems: the engineering approach. In J. Ganoulis, (Ed.), *Water Resources Engineering Risk Assessment*. New York, NY: Springer-Verlag.

Haimes, Y. Y. 1991. Total risk management. *Risk Analysis*. 11(2): 168–71. New York, NY: Plenum Publishing.

Haimes, Y.Y. 1998. *Risk Modeling, Assessment, and Management*. New York, NY: John Wiley and Sons.

Haimes, Y.Y., K. Tarvainen, T. Shima, and J. Thadathil. 1990. *Hierarchical Multiobjective Analysis of Large-Scale Systems*. New York, NY: Hemisphere Publishing.

Haimes, Y.Y., N.C. Matalas, J.H. Lambert, B.A. Jackson, and J.F.R. Fellows. 1998. Reducing the vulnerability of water supply systems to attack. Charlottesville, VA: Center for Risk Management of Engineering Systems, University of Virginia. Submitted to *Journal of Infrastructure Systems*.

HydroSEC. 1990. Kim Petersen (Ed.), *Water Security* (the security newsletter of the water utility industry). Herndon, VA: HydroSEC, Inc. (no longer in existence).

Kaplan, S. 1997. The words of risk analysis. *Risk Analysis* 17(4): 407-418.

Kaplan, S., and J. Garrick. 1981. On the quantitative definition of risk. *Risk Analysis* 1(1): 11-27.

Kumamoto, H., and E.J. Henley. 1996. *Probabilistic Risk Assessment and Management for Engineers and Scientists*, 2nd Ed. New York, NY: IEEE Press.

Kunze, D.R. 1997. Assessing utility threats. *Security Management* (monthly of the American Society for Industrial Security) 41(2): 75-78.

Kuzminski, P., J.S. Eisele, N. Garber, R. Schwing, Y.Y. Haimes, D. Li, and M. Chowdhury. 1994. Improvement of highway safety I: identification of causal factors through fault-tree modeling. *Risk Analysis* 15(3): 293-312.

Lambert, J.H., N. Matalas, C.W. Ling, Y.Y. Haimes, D. Li. 1994. Selection of probability distributions for risk of extreme events. *Risk Analysis* 14(5): 731-742.

Magli, T. 1993. Investigation work begins at Peruca Dam, Croatia. *International Water Power and Dam Construction* 45(4): 15-22.

Matalucci, R. 1998. Personal communication of a course on architectural surety designed by Matalucci at the Department of Civil Engineering, University of New Mexico, and Sandia National Laboratories, Albuquerque, NM.

Ownbey, P.J., F.D. Schaumburg, and P.C. Klingeman. 1988. Ensuring the security of public water supplies. *Journal of the American Water Works Association* 80(2): 30-40.

PCCIP. 1997. *Critical Foundations: Protecting America's Infrastructures.* President's Commission on Critical Infrastructure Protection, Washington, DC: US Government Printing Office.

Roberts, N.H., W.E. Vesely, D.F. Haasl, and F.F. Goldberg. 1981. *Fault Tree Handbook.* US Nuclear Regulatory Commission, Washington, DC: US Government Printing Office, Document NUREG-0492.

Smith, B.L. 1994. *Terrorism in America.* Albany, NY: State University of New York Press.

Risk-Based Decision Making For Dam Safety

Jerry L. Foster, PE[1]

Abstract

The US Army Corps of Engineers (USACE) has been using risk analysis to make investment decisions for several years on proposed major rehabilitation projects which do not involve public safety issues. Efforts are now underway to extend the use of risk analysis to the Dam Safety Assurance Program. This paper presents the procedures proposed for USACE projects and discusses some of the issues which must be resolved if risk analysis is to be used to make safety-related decisions.

Introduction

The USACE has adopted the policy of using risk assessment to make all investment decisions, and guidance has been developed for using risk and reliability analyses for major rehabilitation projects. This policy has been applied primarily to navigation projects and powerhouses which do not involve risk to human life. Efforts are now underway within USACE to explore the possibility of extending risk assessment methods to evaluate projects under the Dam Safety Assurance Program (DSAP). While the USACE development of risk analysis procedures has begun, many policy and procedural issues remain to be resolved. Therefore, the following discussions reflect the author's opinions and should not be interpreted as USACE policy.

All federal and state agencies responsible for the design, construction, operation, or regulation of water resource projects have recognized the need for making sound investment decisions regarding dam safety. Agencies have been searching for a systematic method for prioritizing needed repairs to their dams. At this time, the Federal Coordinating Council for Science and Technology Policy (FCCSTP), Office of the Science and Technology Policy (OSTP), is placing

[1]Structural Engineer, US Army Corps of Engineers, 20 Massachusetts Ave., Washington, DC 20314

placing increasing emphasis on having all federal agencies present their annual programs and budgets using a risk-based approach.

Background

As a dam owner, builder, and operator, the USACE needs a risk management policy to assure that public safety and economic investments are adequately protected. While many USACE structures have not been subjected to their maximum design conditions, 65 of the Corps' 569 dams have been identified as being hydrologically or seismically deficient. In addition, approximately 64% of our dams are over 30 years old and 28% have reached or exceeded their 50-year design life. Many of these older structures need major repair or rehabilitation to ensure their continued safety for future generations. Dam safety modifications are expensive, and decisions to prioritize modifications must balance the risks associated with failure against the current funding limitations. Risk assessment is an important tool for decision makers and engineers. It provides a rational basis for incorporating the event frequency and performance of the dam into decision making, and introduces into the process the impact on the threat to human life of alternative repair schemes. The use of risk analysis for making dam safety decisions, however, has historically been rejected by both decision makers and design engineers. The primary objections are based mainly on the reluctance of policy makers and designers to accept, publicly, the fact that dams fail. Other arguments against risk analysis methods previously proposed revolve around the requirements of these methods to select a dollar value for human life.

Dam safety policy for determining hazard classifications is established in USACE regulations (USACE 1997). The number of lives lost which defines the threshold for high-hazard projects is stated in this policy as one additional life lost due to failure of the structure. The USACE policy for determining hydrologic or seismic deficiencies does not directly include structural performance of the dam under the postulated event, nor does it consider the frequency of the event. In order to determine the level of project hazard, the event and structural failure are assumed to have occurred, and structural stability is independently analyzed to calculate deterministic safety factors that are compared to design criteria. This often results in very conservative assessments of stability and generally predicts failure for unusual and extreme loading conditions.

Federal policy for evaluating spillway capacity is that a dam is considered to meet safety criteria if failure related to hydrologic capacity will result in "no significant increase in downstream hazard" over the hazard that would have existed if the dam had not failed. This policy does not define what constitutes a significant increase in downstream hazard. A base safety condition (BSC) is determined by comparing the loss of life projected for various percentages of the probable maximum flood (PMF) with and without dam failure. The flood expressed as a percentage of the PMF for which "loss of life is not significantly different with and without dam failure conditions" is the BSC. These terms are too vague for dam safety criteria. For example, is the difference between one life lost and five lives lost significant enough to change the PMF percentage to be used as the BSC? Is the loss of 1000 lives due to failure during a PMF with a return period greater than 2000 years more significant than the loss of 10 lives during a failure from a flood with a return period of 50 years?

The policy for determining BSC flood levels also allows the use of economic risk analysis for flood effects above the BSC. The vagueness of the loss-of-life definition used to establish the BSC can lead to using economic risk analysis to make dam safety decisions which will also involve loss of life. This can lead indirectly to establishing a dollar value of human life by allowing alternative levels of downstream protection, with differing loss-of-life potential, to be compared on the basis of economics.

Probabilistic Aspects of Dam Safety Assessments.

The current process requires the deterministic assessment of many parameters which are better defined probabilistically. An example is the population at risk (PAR) and the fatality rate within the PAR. The PAR for a particular failure event can be assumed to be a deterministic parameter. However, the fatality rate must distinguish between total population downstream and the population that would likely be in a life-threatening situation given the extent of prefailure flooding, warning time available, evacuation opportunities, and other factors that might affect the occupancy of the incrementally inundated area at the time the failure occurs. The fatality rate is therefore a variable which should be determined probabilistically, not deterministically.

Many other aspects of the evaluation of a dam for safety are probabilistic in nature. Structural performance under all loading conditions is a function of random variables which include magnitude and distribution of uplift pressures, performance of drainage and grouting systems, material properties, and reservoir levels. The PMF for a particular site is also a function of random variables such as antecedent events, the functioning and operation of gates and outlets, infiltration rates, and maximum precipitation.

Proposed Comparative Risk Assessment for Dam Safety

The use of risk-based dam safety assessments has been proposed for the evaluation of USACE projects, and policy development for the use of these procedures has begun. Under the proposed procedures, the risk would be evaluated based upon the probability that any one life would be lost in the event of a dam failure, i.e., the annual probability of a person dying in a dam failure. This approach allows for the risk associated with the operation of USACE projects to be compared to risk associated with aspects of everyday activities and other federally regulated activities.   Risks are generally classified as either *voluntary* or *involuntary*. Voluntary risks are those due to events under personal control of the persons at risk, such as air travel, auto travel, and the use of machinery or firearms.   Involuntary risks, which are beyond the control of the affected population, are those due to natural events or due to the actions of others.   Examples include hurricanes, lightning, and nuclear accidents.   Society generally accepts a higher level of voluntary risk than of involuntary risk. Risks to downstream populations due to the operation or failure of USACE flood control projects are involuntary.

The risk assessment procedures proposed for dam safety considers two types of losses: *reversible* and *irreversible*. Reversible losses are basically economic, such as property damage, while irreversible losses include loss of life and environmental damage which cannot be mitigated. Dam safety decisions involving reversible losses can be assessed based upon economic risk analyses. However, irreversible losses cannot be based solely on economic considerations. USACE policy establishes the threshold for determining when a dam failure is hazardous to downstream populations as when the failure of the structure will cause one more death than if it had not failed (USACE 1997). This criterion is used to establish hazard ratings for USACE projects and is used herein to define irreversible losses.

The irreversible risk associated with a dam project must consider the frequency of the flood or earthquake event, the probability of structure failure during that event, the total downstream population in the expected resulting flood plain, and the lives expected to be lost in the failure. An event tree can be used to evaluate irreversible risks; however, for the purpose of illustration the following equation is used to express risk:

$$R_t = P_{fatality} * PAR * P_{failure} * P_{event} \tag{1}$$

Where:

$P_{fatality}$ = Probability of life in the inundation zone being lost
             in the event of a dam failure, (fatalities/total lives exposed).

PAR = Total population at risk within the inundation zone
             in the event of a dam failure (total lives exposed/failure).

$P_{failure}$ = Probability of structure failure for the event postulated,
             i.e., PMF, 100-year flood, OBE, MCE, etc. (failures/event).

P_{event} = Frequency of the event (events/year).

The risk determined using the above equation is the product of conditional probabilities for a particular event, and the PAR is also based upon the specific event, i.e., PMF, MCE, or 100-year flood. Therefore, the risk is unique for the event and will be different for other events. The resulting risk is expressed in terms of total annual expected fatality rate (lives/year) associated with the project.

The determination of hazards at USACE projects will be based upon the threat to a single life. If one more life is expected to be lost in a structural failure during the postulated event than if the structure had not failed, the project will be considered a high-hazard project (USACE 1997, Table E-1). While this is consistent with the current USACE policy, the parameters used to determine the expected loss of life are treated probabilistically in the proposed procedures. Using the parameters defined in Eq. (1), the total expected loss of life in a specific dam failure is:

$$\text{Expected loss of life} = \text{LOL} = P_{\text{fatality}} * \text{PAR} \qquad (2)$$

If LOL is determined with and without dam failure, where LOL' is the expected loss of life without a dam failure, then a high-hazard project is one for which

$$\text{LOL-LOL'} > 1 \qquad (3)$$

If the project is determined to be high hazard according to Eq. (3), then the risk analysis would be conducted based upon the risk to human life and not solely upon economic considerations. The annual risk to one life is then:

$$Rp = P_{\text{fatality}} * P_{\text{failure}} * P_{\text{even}} \qquad (4)$$

The risk from Eq. (4) is the annual probability of fatality to each person living in the flood plain. If this value is compared to other similar risks, an evaluation can be made of the level of risk posed to downstream individuals by USACE projects.

Total Risk.

An event-tree analysis can be used to determine the risks to human life for the full range of events expected at the project. The sum of the risks along the various paths of the tree determined in this manner will be the total risk to life posed by the project.

Risk Management

Determining the human and economic risks at flood control projects is a key element in developing an overall risk management policy for a dam owner. Risk management can be accomplished for a specific project, or for the entire flood control and dam safety program, by examining the impact of flood control budget decisions upon each of the parameters defined in Eq. (1). For example: if the threats to human life, determined in Eq. (4), were excessive at some USACE projects due to increases in probabilities of structure failure, additional funding could be concentrated as required on a regional or Corps-wide basis to correct structural deficiencies. The converse is also true: if the risks to human life were much lower at particular projects, then any DSAP proposals for those projects could receive a lower priority. The impact of corporate budget decisions on the parameters in Eq. (4) can be directly quantified, and budget policies can be adjusted to maintain risk exposure at a minimum level. Risk management could therefore be an important tool for USACE decision makers to use in establishing budget priorities while minimizing risk exposure.

Equal Hazard Analysis.

The proposed risk assessment process would limit the total human risk level in Eq. (1) to less than one fatality over the life of the project. This criterion would be used to establish the need for rehabilitation and as the basis for comparing and prioritizing proposed dam safety projects at various sites. Once the need for rehabilitation was established, alternative rehabilitation solutions would be compared based upon the annual risk to individuals within the downstream basin. The level of protection required would be established based upon providing the minimum level of individual risk, and alternatives within that level of protection would be compared based upon an equal-hazard analysis. This method allows for selecting the alternative with the lowest cost from only those alternatives with equal expected fatalities. Human risk is then the primary basis for the decision and a dollar value is not placed on human life. Economic risk analysis should only be used to compare project alternatives with equal loss-of-life projections.

Major Issues to be Addressed by Research

Major issues exist concerning the use of risk analysis to make dam safety decisions. Some of these issues have existed for many years and will require extensive R&D efforts as well as difficult corporate policy decisions. The following quotation explains why risk analysis has been historically rejected as a tool for making dam safety decisions: "Deliberately accepting a recognizable risk to life in the design of a dam simply to reduce the cost of the structure has been generally discredited from an ethical and public welfare standpoint" (USACE 1991). This logic, while meant primarily for the design of new dams, is also applied to the decision-making process for evaluating existing dams. The fallacy of this statement is that there is a risk associated with any project; even the structures designed using the conservative dam-safety criteria currently in effect have a risk of failure. In addition, using conservative criteria to design and evaluate structures does not necessarily result in a conservative or risk-free structure, especially when there is great uncertainty in the assumptions or data used in the design. Risk and reliability analyses quantify the uncertainty and provide decision makers with a realistic view of the consequences associated with the various dam safety alternatives. It is imperative that additional research be conducted to address and resolve these issues and refine analysis methods. The following are major issues which must be resolved if risk analysis is to be used to make dam safety decisions.

USACE will have to make corporate decisions regarding acceptable levels of total risk and involuntary personal risk for flood-control projects. This will require research analyzing the risks associated with USACE flood control projects compared to the risks involved in other similar activities, such as construction of bridges, buildings, or nuclear power plants. The major problem with establishing an acceptable level of risk is that expected loss of life due to a dam failure may be large for a particular event, despite a low total or personal risk associated with the same event. There is a societal aversion to large loss-of life-events, while a small loss of life is generally more acceptable for many events. For example: society is generally alarmed when a plane crashes killing 200 people, but barely acknowledges daily motor vehicle deaths despite the fact that 41,000 people are killed each year in auto accidents and only about 1,100 die in airplane accidents. Policies for acceptable human risk should avoid selecting a specific number of fatalities and should prioritize potential dam safety projects with the goal of reducing human risk to the minimum levels possible within the budget constraints. This can be accomplished by using a risk management policy which examines the risks associated with a project or flood control system and concentrates funding on repairs or improvements which will achieve the greatest reduction in human risk.

One approach for addressing this concern is to use a two-tier system to evaluate risk. Tier 1 (Basic Safety) would require that the annual risk to an individual be less than a specified level, and Tier 2 (Public Trust) would require a decreasing probability of failure as the project fatality potential increases. Potential dam safety projects would then be prioritized with the goal of reducing human risk to the minimum levels possible by concentrating funding on repairs or improvements which will achieve the greatest reduction in human risk.

The use of risk analyses will require that frequencies of occurrence are determined for extreme events such as PMF and MCE. There has been historic reluctance within USACE to establish frequencies for PMF. The frequency for a PMF would have to consider the frequency of the antecedent event (usually an event which fills the reservoir). Therefore, the frequency associated with a PMF will be the combined probability of two extreme events occurring.

Current risk analysis methods for major rehab define structure performance in terms of unsatisfactory conditions, i.e., deflections, cracking, and others. Predicting the probability of structural failure will require determining what constitutes failure. Research will be necessary in the use of analyses which model the nonlinear behavior of a structure at failure using ultimate and nonlinear material strengths.

Risk analyses of alternative dam safety repair schemes will require that a target probability of failure be selected for the design of the alternatives. The target probability of the structure failure should vary depending upon the frequency-of-loading condition under investigation. Care must be taken to assure that only alternatives with equal levels of protection are compared in order to avoid indirectly establishing a value of human life in the analysis. Alternatives with different levels of protection will have different human risk and project costs associated with them, and thus can lead to placing a value on human life. Alternatives with different levels of protection must be compared on the basis of Eqs. (1) and (4).

R&D Efforts Past, Present and Future

To date, there has been a limited effort to develop risk analysis methods in the Risk Analysis For Water Resources Investment (RAWRI) research program. While this effort has clearly demonstrated the benefits of risk analysis research for making decisions regarding flood damage reduction and major rehabilitation, the program has not addressed dam safety. The proposed R&D plan for dam safety would have generic work units in a RAWRI dam safety sub-program focus on the evaluation framework and on basic uncertainty quantifications needed to make program decisions prioritizing studies, repairs, and modifications for USACE projects. A second, and separate, R&D program would be established to develop the technical aspects of risk analysis methods required for evaluating dam safety problems in depth. These procedures would then be used to refine and optimize the alternative repair, warning, and modification plans and designs for each project.

While the research program for dam safety has not been finalized, the objectives of the USACE research and development plan are expected to include the following :

-       to develop new risk analysis methods,
-       to improve the Corps decision-making ability for prioritizing the dam safety program, and develop a dam safety plan for each project using risk-based tools,
-       to demonstrate the use of risk-based tools for making dam safety decisions,
-       to provide assistance in deploying this methodology, and
-       to promote technology transfer throughout the dam safety community by cooperative agreements and cost-sharing efforts with other federal and state agencies and with the private sector.

Research proposed in the new program would focus on the development of procedures for specific aspects of the risk analysis, including:

-       flood frequency relationships and improved hydrologic modeling,
-       improved dam breach analyses and identification of failure mechanisms,

- improved dam monitoring techniques,
- methods for development and evaluation of flood warning systems,
- probabilistic determination of ground motions,
- probabilistic characterization of the site, embankment, and structures,
- reliability of the operating equipment and structures,
- extrapolation of uplift data to higher pool conditions and improved seepage modeling, and
- management and implementation/technology transfer.

Conclusions

Developing and implementing risk analysis methods for making dam safety decisions will enable the Corps to prioritize the inventory of dams requiring engineering investigations and subsequent analyses, and to prioritize funding for critical dam safety repairs or modifications. Risk assessments can also be used to select the optimal alternative plan for protection of public safety, reduction of property damage, and mitigation of environmental damage. This will minimize disruptions of the services provided by USACE projects while maximizing the effectiveness of infrastructure investments.

References

USACE. 1991. *Inflow Design Floods for Dams and Reservoirs.* ER 1110-8-2 (FR), US Army Corps of Engineers, Washington, DC.

USACE. 1997. *Dam Safety Assurance Program.* ER 1110-2-1155, US Army Corps of Engineers, Washington, DC.

Achieving Public Protection with Dam Safety Risk Assessment Practices

Charles Hennig[1], Karl Dise[2], Member, and Bruce Muller[3], Member

Abstract

In 1978, the Bureau of Reclamation implemented a dam safety program in accordance with the Safety of Dams Act of 1978. In the early years of the program, the most significant safety risks to the public were readily apparent to decision makers and were corrected. With many of the most serious dam safety deficiencies corrected, Reclamation has been challenged with identifying and prioritizing future corrective actions in a manner which will provide reasonable improvements in public protection. While Reclamation has previously used risk assessment approaches for the evaluation of potential economic losses, the agency is now implementing regular use of risk assessment to evaluate and prioritize issues involving the personal safety of the public. Two key elements of the implementation of risk assessment methods include agency guidelines for achieving public protection and measures to be employed in identifying estimated risks to the public.

The Dam Safety Challenge

The Bureau of Reclamation is responsible for the safety of 382 high and significant hazard dams in the 17 western states. Approximately 50 percent of this inventory is more than 50 years old. In addition, approximately 90 percent of the inventory was constructed before many of the current state-of-the-art design and construction practices in use today. Reclamation faces significant challenges to ensure that this aging inventory of dams can continue to safely perform beyond their original design intents, which were based on the design practices in use when these

[1]Program Manager, Dam Safety Office, Bureau of Reclamation, PO Box 25007, Denver, CO 80225
[2]Geotechnical Engineer, Technical Service Center, Bureau of Reclamation, PO Box 25007, Denver, CO 80225
[3]Civil Engineer, Technical Service Center, Bureau of Reclamation, PO Box 25007, Denver, CO 80225

structures were built. Integrating risk management principles into dam safety decisions is essential to help focus resources toward those activities that achieve the most effective and efficient risk reduction.

Meeting the Dam Safety Challenge

As structures age, continued safe performance becomes a greater concern. The Bureau places great reliance on recurring and ongoing dam safety activities to detect, intervene, and effectively respond to dam safety incidents. These activities include structural performance monitoring, emergency action planning, operator training, and an aggressive examination program to help identify developing problems and issues that may require additional investigations. A blanket level of risk management, or risk reduction, is provided across the inventory by such recurring and ongoing dam safety activities.

More emphasis is being placed on methods to enhance monitoring practices by identifying site-specific failure modes and customizing monitoring to effectively observe the performance associated with each failure mode. Methods to automate the detection of significant changes in performance, such as increased seepage and changes in reservoir levels, are also being pursued where beneficial. Emergency action plans are being updated to include site-specific indicators of developing problems along with education of downstream officials and testing of the plans. The examination program consists of ongoing visual monitoring, an annual inspection, and periodic and comprehensive examinations which alternate on a three-year basis. The periodic examination is a complete condition inspection and status review conducted every three years by a specialist. These activities are repeated during the comprehensive examination with the additional participation of a senior-level dam design engineer. Additional activities during a comprehensive examination include a complete review of performance monitoring requirements and an assessment of issues that may be affected by current state-of-the-art practices.

Risk assessment practices are also being integrated into the Dam Safety Program to help understand the many uncertainties associated with the continued safe performance of existing dams and their impacts on risk. Risk assessment approaches are intended to be an additional tool that leads to improved decisions by helping to accomplish the following objectives:

- Recognizes all dams have some risk of failure
- Considers all factors contributing to risk
- Identifies the most significant factors influencing risk and uncertainty, which facilitates efficient targeting of additional data and analyses
- Identifies a full range of alternatives to manage risk, including monitoring and other non-structural methods

- Focuses funding and resources toward risk-reduction actions that achieve balanced risk between dams and between failure modes on individual dams
- Establishes stakeholder credibility and due diligence for risk-reduction actions

The Role of Risk Assessment in Dam Safety Practices

Using risk assessment approaches to assess dam safety is not a new idea. The Federal Guidelines for Dam Safety encouraged the development of risk-based approaches to dam safety. These guidelines were implemented for dams regulated by the federal government by a presidential memorandum dated October 4, 1979. Risk assessment practices were initially focused on evaluating the economics of proposed corrective actions. However, their use diminished as experience showed that most dam safety decisions are driven by concerns for the safety of the public. During the past 10 to 15 years, most dam safety deficiencies were relatively obvious. Issues such as active piping did not require extensive investigations to assess the reliability of continued safe dam performance and need for modifications. Dam safety issues today are typically becoming more complex as the continued safe performance of existing dams, especially during extreme earthquake and flood events, are more central problems. Risk assessment practices facilitate the evaluation of complicated risk factors and the influences introduced by associated uncertainties.

The Bureau of Reclamation quantifies risks to public safety based on the expected values of the consequences:

$$\text{Risk} = \text{Estimated Average Annualized Loss of Life}$$
$$= (P_{\text{Load}}) (P_{\text{Failure}}) (P_{\text{Exposure}}) (\text{Consequences})$$

Where:
P_{Load} = Probability of load
P_{Response} = Probability of an adverse response given the load
P_{Exposure} = Probability of being exposed to adverse conditions or consequences
Consequences = Estimated loss of life for the conditions analyzed

For each load category (seismic, flood, and static), risks are evaluated under a full range of loads. An event tree is constructed around the framework of this expression to represent the various failure modes and linked events associated with the full development of each failure mode. Consequences may include economic losses, potential for loss of life, or other adverse consequences associated with uncontrolled releases from a dam.

Why should we, as dam safety professionals, want to assess dam safety in this manner? To help answer this question, let's consider the objective of dam safety. The key dam safety questions are:

- How can an existing dam fail?
- How safe is it?
- How safe is safe enough?
- How can risks be managed?

Dam safety is fundamentally different from dam design. The designer's paradigm uses design standards and safety factors to evaluate the safety of existing dams in relation to a commonly accepted level of conservatism. Traditional standards are mostly intended to establish a level of confidence that a design will result in a successful, load-tested structure, given uncertainties in the loading conditions and capacity of the structure to resist load. However, these standards fall short of providing a systematic mechanism to help evaluate and answer dam safety questions in a way that helps the agency identify reasonable corrective actions.

Extensive use of engineering judgement is required to answer these questions. Decisions based on engineering judgements have always been fundamental to dam safety assessments. This is due primarily to the many material properties and model uncertainties inherent in the investigative methods currently available to assess the performance of existing dams. Existing dams are typically plagued with uncertainties, because they often lack many state-of-the-art features. For new dams, many of these uncertainties are addressed in the design and construction stage, such as removing foundation materials subject to liquefaction, incorporating embankment zoning and filtering to control and filter seepage, and providing structure geometry that ensures linear behavior.

Risk assessment approaches provide a mechanism to quantify this judgement. Quantified judgement not only helps assess the question of how safe, but also permits the probabilities of being wrong or right to be considered in the decision-making process. Quantified judgement also provides a means to prioritize program efforts on the basis of risk between dams as well as failure modes, in order to achieve the most effective risk reduction with available resources and a more balanced approach to dam safety.

Some examples help illustrate this discussion:

Case 1a: After completing traditional analyses to assess the ability of an embankment dam to withstand the maximum credible earthquake, a typical judgement-based conclusion might be that the crest of the dam is not expected to deform more than available freeboard and therefore failure is unlikely. This conclusion provides the decision makers with little information concerning our confidence in the conclusion. It also provides little or no information regarding uncertainties regarding the understanding of loading conditions, dam behavior, or the potential consequences. Combining a low probability of failure with a potentially high load probability and high consequences may lead decision makers to believe that they are overexposed and that the chances of being wrong about the dam response are too great.

Case 1b: Another typical response to this example could also be that crest deformations are likely to exceed available freeboard and cause dam failure. A standards-based conclusion that the dam is unsafe provides the decision maker with little information concerning risk in comparison to other dams in the inventory. Reclamation typically has several dams with issues to address at any given time. An effective dam safety program should provide the information required for prioritizing the needs for corrective action on the basis of overall improvements to public safety.

Case 2: The conclusion of another analysis might be that seepage, although unfiltered, is not likely to cause piping. Such a conclusion does little to help the decision makers understand the level of risk that is being accepted. It also fails to evaluate whether or not the accepted risk is comparable to the risks associated with other potential failure modes. Merely meeting accepted design standards may not portray the importance of particular failure modes when the loads leading to these failure modes have very different probabilities of exceedance.

Traditional standards-based approaches often place substantial emphasis on maximum credible earthquake (MCE) or probable maximum flood (PMF) issues without facilitating, and often concealing, a comprehensive look at all contributing load levels and risk factors. Risk assessment approaches allow all factors that contribute to risk to be considered, which can lead to a better understanding of risk and consequently more effective risk management. Reviewing the components of the risk equation helps to establish a better appreciation for this understanding:

Loading:
Standards-based approaches typically use MCE, PMF, and normal water surface (NWS) for load evaluations. These standards do little to inform the decision maker about exposure to the highest risk events. To effectively manage risks, decisions need to be made with knowledge of the lowest level of loads and associated probabilities that can cause dam failure to initiate. These loads could be very different from the MCE, PMF, or NWS. In addition, large earthquakes and floods in some regions of the west are usually more probable than similar events in other regions. Standards-based approaches do not provide a way to assess these regional impacts on risk to facilitate effective program prioritization.

Structural Response:

Structural response probability under a full range of loads for a given load category (e.g. earthquake, flood, static) is comparable to the safety factor in a standards-based approach. Required safety factors are frequently established somewhat arbitrarily such that a design is adequately conservative. Using an assigned factor of safety criteria does not recognize different levels of data and model uncertainty from site to site and tends to assume that the influencing factors are uniformly understood. Is a 1.0 factor of safety for seismic stability adequate for each site and the range of potential uncertainties that may exist? Could failure still occur? Is the requirement for a 4.0 factor of safety for foundation stability under normal loads unreasonable for a load-tested structure? Would it be more effective to direct resources to higher-risk issues on another dam?

Exposure:

Reservoir levels that create unsafe conditions under various loading scenarios commonly may occur during certain periods each year. This is especially true for Reclamation irrigation storage facilities that typically fill during the spring but are drawn down by the end of the irrigation season. Seismic stability, flood overtopping, or other issues evaluated only under full reservoir conditions would overstate risk. Likewise, it may be common for some reservoirs to operate in flood surcharge, in which case evaluation at normal reservoir levels could understate risk for various issues and possibly even conceal certain safety concerns that arise between normal reservoir elevation and crest of the dam. In addition, exposure to consequences can have seasonal variations that may elevate or reduce risks. Standards-based dam safety assessments do not accommodate variations in risk-exposure factors in a meaningful way and do not facilitate an identification of these factors. The consideration of risk-exposure factors is a natural outcome of risk assessment approaches.

Consequences:

A few individuals at risk 20 miles downstream represents a different level of risk from that of a few individuals who reside at the toe of the dam. A large metropolitan area at the toe of the dam represents an even greater level of risk. Flood-wave travel time affects warning time and time available to evacuate the population at risk, which directly influences the number of fatalities expected during a dam failure. Standards-based dam safety assessments focus only on the structure. This not only ignores the consequence contribution to risk, it also tends to distract the analysis and decision processes from the fact that managing/minimizing adverse consequences is the essence of dam safety. Risk assessment approaches illuminate factors that keep objectives focused on methods to effectively and efficiently minimize consequences.

Traditional analysis and investigation techniques remain an essential component of dam safety assessments. When combined with risk assessment approaches, a better understanding of risks is achieved, which facilitates good decision making. Traditional techniques are essential for providing an understanding of structural behavior and the potential limitations of model uncertainties and material property variability. The traditional analysis results become a significant source for structure-response probability estimates in the framework of a risk assessment.

Public Protection Guidelines

Interim guidelines for achieving dam safety public protection (USBR 1997) have been established so that estimated risks can be measured against the justification for risk-reduction actions. "Public protection" terminology is used instead of "acceptable risk", because the program emphasis is on achieving public protection. "Acceptable risk" criteria tends to establish a mindset that no risk-reduction actions should be considered if risks are below the criteria. However, prudent program practices should thoroughly understand the nature of risks and always look for opportunities to efficiently reduce them in a cost-effective manner. "Acceptable risk" terminology also creates an attitude of callous insensitivity to public safety; it does not recognize the fact that all dams have some risk of failure no matter how well designed or constructed. Guidelines are used, in lieu of specific criteria, so that site-specific influences and conditions not easily or reliably represented in a risk assessment framework can be considered in the decision. These guidelines do allow deviations, but require an understanding of the basis for the deviation.

Figure 1 represents *Tier I Guidelines* and focuses on potential loss-of-life considerations. This figure is used to plot the estimated expected annualized loss of life for each load case, evaluated as follows:

A. Estimated Average Annual Loss of Life > .01:

Risk is typically considered elevated to the level that there is strong justification to take actions to reduce risk for continued long-term operations. In addition, risk-reduction actions should be considered during the interim until the permanent modifications can be designed and implemented. Interim actions could consist of operating under reduced reservoir levels, enhanced monitoring, enhanced emergency preparedness, and/or various structural measures that reduce estimated risks below .01.

B. Estimated Average Annual Loss of Life between .001 and .01:
    Strong justification to reduce risk for continued long-term operations. There is not strong justification for interim risk reduction actions provided modifications could be implemented within approximately five years. However, easy opportunities to better manage risks during the interim should not be overlooked.

C. Estimated Average Annual Loss of Life < .001
    The justification to implement risk-reduction action diminishes as estimated risks are increasingly smaller than .001. Corrective action costs and the associated amount of risk reduction that could be achieved are factors that influence decisions. Opportunity costs for

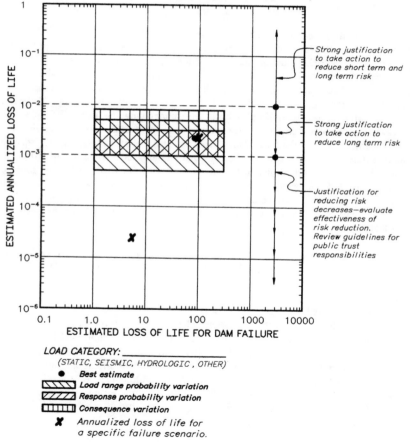

Figure 1. Tier 1 Guidelines (Loss of Life)

taking actions in this range are also considered in terms of available agency resources that would be forgone for efforts on higher-risk facilities or failure modes. In addition, other water-resource management issues begin to play a more significant role in the decision. Decisions to take no risk-reduction actions are not considered permanent. Issues and associated risks are revisited on a minimum six-year recurring basis, recognizing that risk factors and agency priorities are subject to change.

Establishing .001 as the zone of differentiation between strong and reduced justification for taking risk-reduction actions creates a sliding level of protection that is proportional to consequences. This is best illustrated by examining the risk equation in the following simplified expression:

Annual Life Loss = Annual Event Probability x Loss of Life

.001 = 1/1,000 x 1 lost life

} 10 times safer design

.001 = 1/10,000 x 10 lost lives

} 10 times safer design

.001 = 1/100,000 x 100 lost lives

This relationship shows that with greater consequences, a more remote design event is required to achieve an adequate level of public protection. This is consistent with societal values which view single events that cause high numbers of lost lives as long-remembered national tragedies. There is significant public aversion to single, high-consequence events, and the public expects a high degree of protection from such events. With low-consequence events, several other factors need to enter into the decision-making process:

1. When loss of life is low, a small population can be exposed to events having relatively high probabilities. Risks become similar to other societal risks such as auto accidents and disease. However, introducing dam-failure risks could significantly contribute to the overall life risks to these individuals.

2. The greater the inventory of dams and the time of exposure, the more likely it becomes that the agency will experience a dam failure as shown in Figure 2. For example, assuming a binomial distribution and allowing 10 dams within the Reclamation inventory of 382 dams to have an average annual failure probability of 1/1,000 yields a 40 percent chance of failure within the next 50 years from just these 10 dams. For 50 dams, the chance of experiencing a dam failure increases to greater than 90 percent. Once a dam failure occurs, public trust is compromised and the public will expect more severe and potentially

more costly protection. In addition, a high level of national safety and stewardship of public assets is expected of an agency entrusted to manage a large inventory of dams.

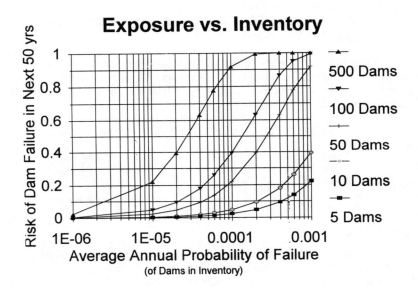

Figure 2. Risk Exposure at Multiple Dams

3. The estimated expected average annual loss of life is based on the overall average risk exposure of the population in question. This does not consider that some individuals within the dam failure inundation zone have a greater exposure to the dam failure than others.

Figure 3 represents the *Tier 2 Guideline*, which is intended to help balance these concerns. Tier 2 establishes the justification for making structural modifications to limit annual failure probabilities to 1/10,000. This ensures a higher degree of protection to small populations and critically exposed individuals than would be potentially provided under Tier 1 Guidelines. It also enhances public trust by being proactive in maintaining the developed water resources in the western states. Taking actions to improve dams that are not equipped with state-of-the-art features to improve their safety is prudent in maintaining our aging infrastructure and public investments. The chance of experiencing a dam failure within 50 years from 10 dams each having an annual failure probability of 1/10,000 is reduced to 5 percent

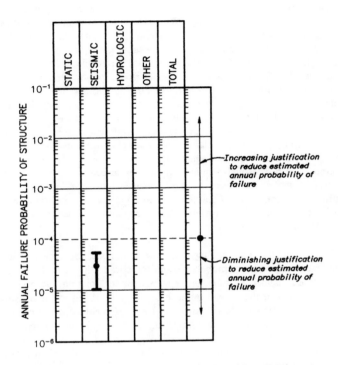

Figure 3. Tier 2 Guidelines (Failure Event Probability)

from the 40 percent chance in the previous discussion. The 50-dam example is reduced from a greater than 90 percent chance of failure down to a 21 percent chance.

Risk Assessment Methodology

In order to apply risk assessment methods to the Reclamation inventory of dams for the purpose of dam safety decision making, it is important that there be a degree of consistency in the methods used for assessing risk. Following two years of applying such risk assessment methods within Reclamation, a document (USBR 1997) is currently being prepared to identify general methods to be used in performing risk assessments. The objective of the document is to gain consistency in applying risk assessment methods with an appropriate level of effort, so that meaningful risk-based information can be incorporated into dam safety decisions.

Some of the issues considered in selecting risk assessment methods included scalability, flexibility, and reliability. With the Reclamation inventory of 382 storage dams, the risk assessment methods must be applicable to a wide variety of dam types, heights, reservoir sizes, and conditions. The same methods that are used to analyze Grand Coulee and Hoover Dams should also be applicable to small diversion dams when the levels of effort are scaled appropriately. The methods should also be flexible enough to allow unique and site-specific conditions to be evaluated within a risk context. Flexibility is also required to allow new developments to be incorporated as Reclamation continues to learn about the use of risk-based information in decision-making processes. In order to achieve maximum risk reduction throughout the Reclamation inventory, the results from each risk assessment should be evaluated against a common basis. Reclamation considers the risk estimates to be reasonably reliable if an internal peer review shows that loading conditions, structural responses, and consequences have been adequately addressed. While no two risk assessment teams would arrive at exactly the same values of risk, the goal is to present sufficient information such that any differences in estimates do not alter the decisions to be made.

Much of the methodology document focuses on recommendations for applying basic principles of probability and statistics to the case of a particular dam. However, there have been three areas in which Reclamation has needed to select approaches which fit the overall purpose of risk assessment in the decision-making process. These areas include selecting the risk assessment team members, estimating load and structural response probabilities, and dealing with uncertainty.

There are many good arguments for a variety of risk assessment team compositions. Many people believe that the credibility of the values determined in the risk assessment is enhanced by the use of world-renowned consultants on the team. Others believe that risk assessments should be performed by teams of individuals who are already familiar with the dam and can directly contribute to the understanding of its behavior. At Reclamation, the core of the risk assessment team is the group of technicians, engineers, and geologists who have an ongoing responsibility for following the behavior and condition of the dam. Since these individuals are generally not trained risk assessment professionals, the team is provided with a facilitator to guide them through the risk assessment process. When there are very sensitive issues involved, it is common practice to involve independent industry consultants either as members of the team or by presenting results to them for review. This approach helps ensure that the risk information provided to decision makers is based on the best collective information available.

From a public protection perspective, the water resources industry has been fortunate that dam failures are unusual events. However, with a relatively limited number of failures, it is difficult to develop meaningful frequency relationships for the failure rates of dams due to specific failure modes. As a result, Reclamation's experience in the past two years is that the probability estimates developed by the risk assessment team can most realistically be thought of as a degree of belief in the annual probability of failure for a given dam under the given conditions. By accepting probabilities based on this "degree of belief" philosophy, risk assessment teams have been able to consider information from a wide variety of sources when developing probability estimates. They can combine the knowledge gained from data collection and analysis, historical failure rates, failure case studies, understanding of physical processes, and understanding of structural behavior. With this approach, the team can make a judgement regarding the expected failure probability of a dam for a given failure mode. Considering data from these multiple sources provides the team members with an understanding of the load conditions that would be required to allow a failure mode to develop. Reclamation recognizes that these probability estimates are not perfectly accurate and may change as additional knowledge becomes available. However, these estimates represent the best risk-based information available for a given dam at a given time.

From a decision-making perspective, risk-based information provides a means for dealing with the uncertainties of managing a water storage facility. When presenting risk-based information to decision makers, it is important to provide a measure of the uncertainties associated with the estimated risk so that the risks can be considered in light of other factors having a bearing on the decision (cost, environmental, social, etc.). While uncertainty can be addressed in a variety of ways, Reclamation has chosen to address it through sensitivity analysis of the estimated risk results.

Application to Dam Safety Decisions

The objective of implementing risk assessment methods in the Reclamation Dam Safety Program has been to improve organizational effectiveness in achieving risk reduction in the existing inventory of dams. The traditional means of achieving this objective has been to determine which dams are unsafe and then implement modifications at those dams. Implementation of risk assessment has allowed Reclamation to go beyond the question of whether or not a dam is safe to incorporate the concept of risk assessment into all phases of the Reclamation dam safety process. Following dam inspections, which occur every three years for each dam, a brief assessment of risks is conducted to determine if any need to be addressed in more detail. Through the use of risk assessment, it is possible to determine if the risk at a particular dam is significant and its relative priority with respect to risks at other Reclamation dams. When additional investigations are required, risk assessment results assist in developing an investigation program focused on the data and analysis with the greatest potential for risk reduction. When corrective actions are determined to be necessary, risk assessment results can be used to guide the development of alternatives that most effectively reduce the risk at a dam. In these ways, risk assessment has become a valuable tool for evaluating and correcting dam safety concerns as an integral part of the Bureau of Reclamation's mission to manage water resources in the western states.

References

USBR. 1997. *Guidelines for Achieving Public Protection in Dam Safety Decision Making.* Interim Guidelines, US Bureau of Reclamation, US Department of Interior, Denver, CO.

USBR. 1997. *Risk Assessment Methods for Dam Safety Decision Making.* Draft, US Bureau of Reclamation, US Department of Interior, Denver, CO.

Toward a Risk-Based Assessment of Shallow-Draft Navigation Investments

Wesley W. Walker[1]

Abstract

Sixty Corps of Engineers' locks and dams on the Ohio River and its major tributaries create some 2,800 miles of waterway navigable by shallow-draft vessels. Major manufacturing and energy-related industries rely on this transportation network. The Corps of Engineers uses system modeling techniques to estimate the national economic benefits of the Ohio River system of waterways. System analysis as practiced in the Ohio River Region of the Corps' Lakes and Rivers Division depends upon system-wide databases. These include traffic projections, lock capacity estimates, and transportation rate and cost estimates, all of which are primary inputs into a static system model capable of solving for equilibrium levels of traffic. The basics of this systems approach were developed in the late 1970s and have been modified over time in an effort to better reflect the benefits of a reliable system of locks and dams. Reliability models are being developed by the Corps in order to predict the performance of individual lock components over time horizons that extend 50 years and beyond. This paper presents a concept for incorporating the results of these component reliability models into the system equilibrium models used to estimate the benefits and costs of performance-enhancing lock improvements.

Introduction

The Ohio river basin, that area drained by the Ohio River and its tributaries, is home to 25 million people. Pittsburgh, Cincinnati, Columbus, Indianapolis, Louisville, and Nashville are the region's largest cities. Many, if not most, of the

---

[1]Regional Economist, Ohio River Region Navigation Planning Center, Huntington District, US Army Corps of Engineers, 502 Eighth Street, Huntington, WV 25701-9020

basin's residents are only vaguely aware of the commerce that moves on the Ohio river basin's system of navigable waterways, made possible by a series of 60 lock and dam projects. The 260 million tons of commodities carried by barges on these rivers each year are the products of coal mines, petroleum refineries, stone quarries, cement plants, and farmers, and the raw materials for construction companies, steel mills, electric utilities, paper plants, aluminum manufacturers, and chemical companies, the foundations of the region's economy.

River transportation revolves around the movement of coal and other bulk commodities and helps to keep their transportation rates low. Therein lies the direct economic benefit of inland navigation. Bulk commodities are most efficiently carried by water (Maritime Administration 1994). Inland navigation extends this efficiency deep into the interior of the North American continent. The availability of this form of transportation, along with the availability of rich deposits of coal (approximately 70 billion tons of demonstrated reserves), have made Pittsburgh, Pennsylvania and Huntington, West Virginia, the second- and fourth-largest coal ports, respectively, in the United States — which is itself the second-largest coal exporter in the world (Energy Information Administration 1996; Water Resources Support Center 1995).

Of course, most of the basin's coal is used by domestic markets — primarily the electric utility industry. Over the last 30 years, much of the region's electricity-generating capacity has moved away from small streams and cities to the Ohio River and its system of navigable rivers. Utility companies were drawn there by dependable supplies of cooling water and access to the low-cost transportation afforded by Ohio River System (ORS) waterways. Electric utilities are estimated to save almost $1 billion annually in transportation costs, a contributing factor to their ability to keep electricity prices at levels significantly below those in other areas of the country.

Estimating System Benefits

Water resource agencies came to a "mutual understanding" regarding the guidelines for estimating waterway benefits. This understanding was recorded in the Federal Inter-Agency River Basin Committee's *Proposed Practices for Economic Analysis of River Basin Projects* in May 1950 (US Senate). In that report, referred to as the Green Book, navigation benefits were identified as the difference between the total "...cost of transportation by an alternative means and the non-project or associated cost of transportation by waterway." In arriving at these benefits, transportation savings are estimated for each unique origin, destination, and commodity movement in the system. Each movement's savings are the difference between the least-costly alternate mode (usually rail or truck) and the existing water routing. Annual waterway system benefits are equal to the product of the volume of traffic moving on the waterway and the transportation savings for each ton of that traffic. In 1995 for example, 260 million tons of traffic moved on the ORS at an average transportation rate savings of $9.00 per ton, making system benefits an estimated $2.3 billion. Projecting system benefits necessarily involves being able to project future traffic levels and transportation rate savings for each movement.

System savings have generally grown over time. Traffic growth has averaged around 3.0 percent annually since 1950, suggesting that benefits must be increasing. However, this very growth can cause average rate savings to decline. At locks too small to handle higher traffic volumes efficiently, congestion leads to a degradation in service (reflected in higher transit times) which erodes rate savings. Traffic-related service degradation has been the primary focus of lock improvement studies over the years, but reduced performance can be traced to causes other than increasing traffic volumes.

Aging projects and heavy usage cause serious reliability concerns for the Corps. In response to these concerns, the Corps pursues aggressive maintenance policies designed to avoid failures of major lock components and the lengthy lock closures they involve. Closures of the large main chambers are especially serious as this necessitates re-routing traffic to the smaller, less efficient auxiliary chamber. During main chamber closures, the typical-size Ohio River tow capable of transiting a main chamber in one 60-minute operation must move through the small auxiliary lock chamber in two operations lasting about 150 minutes. In this situation, where the smaller auxiliary chamber carries the whole of the lock's traffic, lock performance is severely degraded. Service disruptions of this type result in longer transit times, higher waterway transportation costs, and lower transportation-rate savings per ton for those shippers relying on the affected lock.

One way the Corps can minimize these service disruptions is by constructing a larger auxiliary lock. Traffic-delay curves can be used to represent the effect of this solution to degraded service. Figure 1 illustrates a typical traffic-delay relationship, which can be modeled with the use of discrete event simulation. The availability of an auxiliary lock the same size as the main chamber shifts the curve to the right, indicating that two large lock chambers can process the same amount of traffic at lower levels of delay than can the existing lock configuration of one large and one small chamber. Of course, all investment decisions must be economically justified. Do the benefits of a construction investment outweigh the costs? Are the net benefits (benefits minus costs) of the construction investment greater than the net benefits of the project in the absence of the construction investment?

There is an established framework for answering these questions, which is described in the Principles and Guidelines, the latest regulatory successor to the Green Book (US Water Resources Council 1983). Central to this framework is the identification of two future scenarios, or conditions. The future condition at the lock

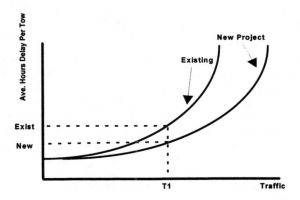

Figure 1. Traffic-Delay Curves, Existing and New Project

without a new construction investment is referred to as the *Without Project Condition*, and the future condition with new construction investment is referred to as the *With Project Condition*.

A key feature of the Without Project Condition for ORS locks is age- and usage-related structural deterioration, a condition which can lead to frequent and/or lengthy closures. Cyclical maintenance and planned component replacements are designed to avoid these types of closures. However, maintenance and repair actions themselves cause closures and lower system performance, often for extended periods of time. These planned closures, proxies for deteriorating structural conditions, must be accounted for in estimating the system's projected benefit stream in the Without Project Condition. While the With Project Condition also can be expected to contain periodic, though less frequent, closures for maintenance, major structural concerns that existed for the old lock will have been eliminated with the construction of a new project. Benefit streams from the With Project investment can be expected to reflect higher levels of system performance due to the relative infrequency of service disruptions. Economic analyses of these competing future conditions seek to estimate the 50-year stream of benefits and costs associated with the respective future.

Current Models and Methods

The discussion above alludes to the key pieces of information used in formulating an economic analysis of alternative waterway investments. This analysis is referred to as a cost-benefit analysis, and the investment with the greatest net benefit (benefits minus costs) is the National Economic Development (NED) plan. Half of the analysis is the cost estimate itself, which is provided by the Corps' project designers. The other half of the analysis is the benefit estimation, a product of systems analysis.

Figure 2 identifies the key inputs to a system benefit analysis: waterway traffic-demand projections, lock capacity estimates, tow operation costs, transportation rates, and lock closure schedules. Each data set represents a considerable amount of work in and of itself. *Traffic demands* indicate unconstrained demands for lock service, *lock capacities* describe the likely performance of the locks given expected fleet configurations, *tow operation costs* allow hours of delay to be expressed as a dollar figure, the difference in *transportation rates* indicates each movement's initial rate savings, and the *lock closure schedule* represents condition-driven lock availability throughout the 50-year project life.

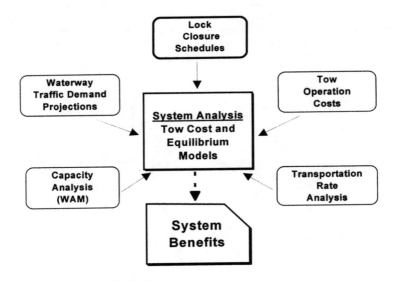

Figure 2. System Modeling Schematic

The model at the heart of the system benefit estimation schematic in Figure 2 is really two separate models — the tow cost model (TCM) and the equilibrium model (EQ). The TCM estimates transit times and the associated trip costs for each unique origin-to-destination commodity movement on ORS waterways for a current year. The EQ model escalates tonnage through the system as indicated by the waterway traffic projections, causing increasing delays at locks experiencing traffic increases through time. As delays and delay costs increase, these costs are used to erode each movement's initial, or base-rate, savings per ton. If the base-rate savings amount is eroded to the point where the waterway routing is more costly than the alternate overland routing, it is assumed that the movement diverts to the alternate mode. This movement is removed from the list of system shipments, and its benefits are no longer included in the estimate of waterway system benefits. The EQ is at solution when it has found that combination of movements with positive rate savings which maximize system utilization in terms of tons accommodated.

Figure 3 depicts an abbreviated system model network. In this network each lock has its own traffic-delay curve. Using the origin and destination ports indicated in the shipment-demand file for a given year, the model calculates the tonnage at each lock and the average delay associated with that level of traffic. It then re-computes the travel time and costs for each system movement individually. These additional, delay-generated transit costs are deducted from the movement's initial rate savings. This new rate-savings figure is multiplied by the movement's annual tonnage to estimate the annual benefits for the movement. Overall, system benefits are probably higher because traffic is higher, but the average rate savings per ton for each movement has been eroded due to increasing traffic congestion. The sum of all

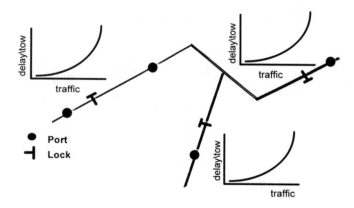

Figure 3. System Modeling Network

re-computed benefits (the product of tons and rate savings per ton) represents the system benefits for the year being modeled. The EQ model is run for each year in the period of analysis, making available a 50-year stream of system benefits.

In the last twenty years, major model modifications have revolved around efforts to better represent the performance of the system's locks. Queuing theory and mathematical representations used in early system models have given way to discrete event simulations of lock performance. New databases were put in place by the Corps to collect the detailed timing data required for lock capacity simulations, and considerable efforts went into developing a credible vessel fleet database. Simulations of this type yield capacity at locks in terms of either an annual throughput capability or a traffic-delay relationship, given the number and dimensions of chambers, the vessel fleet configuration, the lock's availability profile, and the lock operating policy in use.

Not surprisingly, lock availability greatly affects the entire system's performance. Some availability variables — relatively short-duration events such as accidents, weather, and high water — can readily be incorporated into the lock capacity simulation using probabilities based upon historic data. Relatively longer closure events — closures associated with a lock's structural reliability — are rare, and for this reason the frequency and duration are difficult to predict. In the absence of historic data, structural analyses and performance standards have been combined with engineering judgment to predict closure events. The economic consequences of these closures have been estimated by accounting for the cost of repairs and increased towing industry transit costs. These closure-induced delay costs also erode the benefits of the system of locks and dams.

Initially, methods for incorporating structural reliability relied on the use of composite benefit streams. In short, two traffic-delay curves were simulated for each lock site, one for the current lock configuration with both chambers open and one for the lock with only one chamber open. These two curves were inputs into separate 50-year system equilibrium model runs — one for a system with one chamber open and the other for a system with both chambers open. Greater delays occur in the one-chamber system, so system benefits are lower. A composite benefit stream is built from these two 50-year streams. The composite benefits for any given year are the weighted sum of the benefits of a one-chamber and of a two-chamber system, where the weight is the number of days in the year each is in effect. Figure 4 shows what the composite benefit streams might look like for the Without and the With Project Conditions. The area that lies between these two streams represents the incremental benefits of the With Project plan. The greater this area, the greater the incremental benefits associated with the investment in new construction.

Recent refinements in the methodology have centered around more accurately representing the effects of service disruptions on benefits. Traffic-delay simulations are now made at each lock site for a series of closure durations. This is done by adding closure days to the lock availability profile. Annual simulations are made

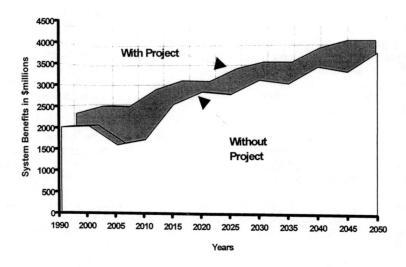

Figure 4. With and Without Project Investment Benefit Streams

with 3 days of reliability-related loss of lock availability, 5 days of loss, 15 days of loss, and so on for each lock site. Closure events are still scheduled externally to the model, but the traffic-delay curve selected by the model for any given year corresponds to the days of closure scheduled for that year.

Emerging Models and Methods

Further refinements and modifications are necessary in order to internalize closure events and move away from strictly deterministic expressions. Much of the subsequent discussion is the conceptual framework for a risk-based assessment of waterway investments. Many of the models referred to are in developmental stages, having moved through early, less-sophisticated forms on their way to fully-developed system treatments. The methodology borrows heavily from techniques used in other investment arenas, such as flood control and hydropower, and really represents the application of a tested technique to a new problem of dramatically larger scope. Major rehabilitation guidance issued by the Corps' Operations and Readiness division was the primary instrument for implementing more stringent analytical requirements in assessing major rehabilitation investments (US Army Corps of Engineers 1995). Because major rehabilitation work is a prominent feature of most Without Project conditions, this guidance has affected the analytical framework for new construction analyses.

Major rehabilitation investments often involve replacing a number of worn or aging major components (usually lock gates, walls, culverts, valves, electrical systems, hydraulic systems, and other critical features). The major rehabilitation guidance requires economic justification for any major component replacement, with this justification evaluating replacement against a condition where the component is allowed to fail before it is fixed — the base condition. Central to this analysis is the ability to both predict when any given component can be expected to fail and to identify the consequences of a failure. Current standards-based approaches to structural reliability assessments are able to identify a component in failure mode, but are not designed to predict when a component is likely to fail.

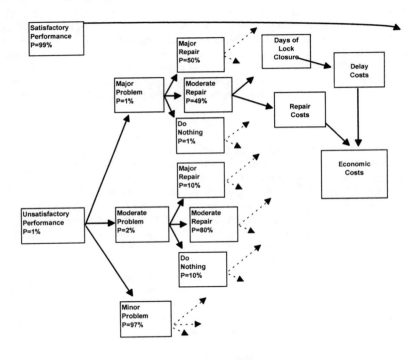

Figure 5. Component-Reliability Event Tree

As a result, engineering reliability analyses are moving away from standards-based descriptions of component reliability to probability-based descriptions, especially for major components. Hazard functions are estimated which describe the probability of unsatisfactory performance (PUP) for each major component. Estimating these hazard functions requires the development of databases on component performance over time and at different levels of utilization. This methodology allows for differing levels of unsatisfactory performance, each with an associated consequence in terms of repair costs and amount of time the lock chamber is closed for component replacement or repair. The final product of the engineering analysis is a component-specific event tree displaying PUPs, level of unsatisfactory performance, the level of response, repair costs, and length of lock chamber closure (see Figure 5).

Each event tree is specific to and bounded by a particular maintenance policy, a particular component, and a particular lock. Thousands of iterations are made through the event tree in a Monte Carlo simulation of the component's performance, yielding a distribution of expected days of closure and expected repair costs due to unsatisfactory performance. Simulations are made for each component and for each year in the 50-year life of a project. If a component is replaced, PUPs are reset to a new or like-new value. Expected closure days are linked to the EQ model, which uses hourly tow costs and traffic-delay relations to estimate expected closure costs associated with unsatisfactory performance. In effect, the event tree and Monte Carlo simulation provide a probability-based schedule of closures that replaces the deterministic schedule used in the current modeling framework (refer to Figure 2).

Conclusion

The Ohio River Region of the Corps of Engineers has been incorporating reliability into its analyses of new construction investments at inland navigation projects in one way or another for the last fifteen years or so. These assessments were largely based upon engineering judgment; attempts at probability-based assessments have only begun in the past few years. Ohio River Region economists first started applying probability-based variations of this technique in the London Lock Replacement report (Huntington District, US Army Corps of Engineers 1996). Later work attempted to link more closely component performance and chamber closures to the EQ model. The latest efforts are being carried out by a team of analysts from Oak Ridge National Laboratory. This work is described in a July 30, 1997 report to the Transportation Research Board (Bronzini et al. 1997). The goal of these evolutionary efforts is to devise a tool that combines the ability to make probability-based assessments of component performance with the capability of analyzing alternative investment and maintenance strategies.

Benefit-cost analysis involving structures nearing the end of their 50-year project life examines the trade-offs between extending the life of the old structure and building a new one. The rehabilitated old project costs less than a new structure, but the act of rehabilitating the lock causes major service disruptions that cost the towing industry and the shippers that rely on this service dearly. Taking reliability into account — taking into account the diminished availability of the main lock chamber — results in a significantly diminished and more accurate representation of the navigation system benefit stream. Accurate representations ultimately lead to an improved ability to express the value of a reliable shallow-draft navigation system.

References

Bronzini, M.S., T.R. Curlee, P.N. Leiby, F. Southworth, and M.S. Summers. 1997. *Ohio River Navigation Investment Model Requirements and Model Design.* Paper submitted to Transportation Research Board. Oak Ridge National Laboratory, Oak Ridge, TN.

Energy Information Administration. August 1996. *US Coal Reserves: A Review and Update.* US Department of Energy, Washington, DC.

Huntington District, US Army Corps of Engineers. 1996. *Kanawha River Navigation Study: London Lock Replacement Interim Feasibility Report.* US Army Corps of Engineers, Huntington District, Huntington, WV.

Maritime Administration. 1994. *Environmental Advantages of Inland Barge Transportation.* US Department of Transportation, Washington, DC.

US Army Corps of Engineers. 1995. *Guidance for Major Rehabilitation Evaluation Reports, Fiscal Year 1998.* Department of the Army, US Army Corps of Engineers, Washington, DC.

US Senate. 1950. *A Report to the Federal Interagency River Basin Committee.* US Senate Subcommittee on Benefits and Costs, Washington, DC.

US Water Resources Council. 1983. *Economic and Environmental Principles and Guidelines for Water and Related Land Resources Implementation Studies.* Superintendent of Documents, US Government Printing Office, Washington, DC.

Water Resources Support Center. 1995. *Waterborne Commerce of the United States, Part 2: Waterways and Harbors, Gulf Coast, Mississippi River System, and Antilles.* US Army Engineer District, New Orleans, LA.

A Systems Approach to the Optimal Safety Level of Connecting Water Barriers in a Sea-Lake Environment. Case Study: The Afsluitdijk Dam, The Netherlands

Alex Roos[1], Frank den Heijer[2], and Pieter van Gelder[3]

Abstract

A systems approach will be developed to determine the optimal safety level of connecting water-barriers in a sea-lake environment. Connecting water-barriers are large dams locking a sea arm or a river branch from the influence of the sea or another river. These dams do not directly protect land or people from flooding, but indirectly reduce the water levels in the enclosed water body. Because of the influence on the water body behind the dam, the water-connecting barrier is an essential part of the flood protection system. The safety level of the water-connecting barrier is the subject of this paper. It will be analyzed in a systems approach with an economic optimization criterion. The approach will be applied on the case study of the Afsluitdijk dam. The Afsluitdijk was built in 1932 to close the former Zuiderzee, a sea arm in the northern part of the Netherlands. After closure the Zuiderzee was called Lake IJssel. After the 1953 storm-surge disaster, several more dams were built to shorten the Dutch coastline. An improvement program was initiated for all the river-, sea- and water-connecting barriers in the Netherlands. Forced by the 1993 and 1995 floods of the Meuse and Rhine rivers, the improvement program must be completed before the year 2000. The optimal safety level of the Afsluitdijk is the subject of this research. No consistent philosophy for connecting water-barriers has yet been established as of this writing. A model has been built to estimate the safety of the dikes around Lake IJssel in relation to the safety level of the Afsluitdijk. With this model the dominant factors can be determined. Finally, a very coarse risk-based optimization has been carried out.

[1]Ministry of Transport, Public Works and Water Management, Road and Hydraulic Engineering Division, PO Box 5044, 2600 GA Delft, The Netherlands
[2]Delft Hydraulics, PO Box 177, 2600 MH Delft, The Netherlands
[3]Delft University of Technology, Faculty of Civil Engineering, Stevinweg 1, 2628 CN Delft, The Netherlands

Introduction

In 1953, a winter depression combined with an exceptionally high spring tide caused a storm surge on the North Sea, pushing water levels to record heights. Dikes failed in several places, especially in the southwestern part of the Netherlands. Immediately after the 1953 disaster, the Dutch government established the Delta Committee. Based on its recommendations (1960), the coastline of Southwest Holland was shortened considerably and a more scientific approach for the design of flood defenses was implemented. For many areas, the shortening of the coastline by the construction of closure dams largely eliminated the threat of the North Sea.

Based on the recommendations of the Delta Committee, large parts of the Dutch coastal defenses had to be improved. The Afsluitdijk, a dam connecting North Holland to Friesland (see Figure 1), also required improvement. In 1932 the Afsluitdijk was built to lock a former sea-arm from the North Sea. The Afsluitdijk is a connecting water-barrier with a length of 32 kilometers. This dam has an inner core of clay and is covered with concrete blocks and grass. A four-lane highway and two lock complexes were constructed. Because the Afsluitdijk was not considered to be a part of the system of dams directly protecting land or people, its improvement had a relatively low priority. After the high river floods of 1993 and 1995 in the Meuse and Rhine rivers, however, the government stated that the entire dike improvement program had to be ready before 2000, including the Afsluitdijk dam.

Because of increased knowledge and changing philosophy relating to water defenses, the question arises whether the Afsluitdijk has to be improved at all. In the new philosophy, a systems approach to flood protection will be used. This paper focuses on the approach to finding dominating aspects of the system and determining an optimal safety level. This approach consists of several steps: ·
• Set the safety concept.
• Describe the system, in particular the most dominating aspects.
• Carry out probabilistic calculations which determine the safety of the system.
• Calculate the cost of required improvements of the dikes.
• Start an optimization procedure to minimize the total costs.
Based on these steps, the optimal safety level can be determined from a technical point of view. This paper discusses these points and presents the first results.

Safety Philosophy Now and in the Future

The present safety standards of the Dutch dikes and other flood defenses are expressed as frequencies of water levels and associated wave impacts that every dike section must be able to resist. These safety standards were set for coastal areas by the Delta Committee after the flood of 1953 in the southwest of the Netherlands, and depend partly on the consequences of flooding. Since that time the safety philosophy has been developed further. The Netherlands were divided into more than 50 polders, called dike ring areas, each protected by a closed system of water defenses and high-lying grounds. Four classes of dike ring areas were distinguished in terms of water level exceedance frequencies: 1/1250,1/2000,1/4000 and 1/10000 per year.

Figure 1. Map of the Area Influenced by the Safety Level of the Afsluitdijk, Including the Study Locations

The Delta Committee derived these classes from the exceedance frequencies by considering economics, and thus they depend indirectly on the consequences of flooding.

The safety standards for connecting water defenses are expressed as a water level frequency as well. However, they are not based on the consequences of flooding. They are in most cases equal to the most stringent safety standards of the connected dike ring areas. For the Afsluitdijk, however, the Delta Committee applied an economic reduction to the safety standard, based on a qualitative estimate of the consequences of flooding. This resulted in a reduction of the design water level, corresponding to a water level frequency of 1/1430 per year.

The Ministry of Transport, Public Works, and Water Management is considering enhancing its flood protection policy. In the future, a flooding risk concept is foreseen (see Van Agthoven et al. 1997) in which both flooding probabilities and the consequences of flooding are taken into account. In the present Dutch Flood Protection Act an evolution from safety standards in terms of water levels towards safety standards in terms of flooding probabilities has been included.

The transition from water level frequencies towards flooding probabilities is not meant to change the current safety level, which seems to be generally accepted and should be maintained. However, a comparison with other types of risk or a weighted distribution of safety expenses to society is difficult when the risks of flooding are not expressed in the same terms as other types of risk.

Economic Optimization of the Dike Height

Taking into account the cost of dike building, of the material losses when a dike breach occurs, and of the frequency distribution of different sea levels, the optimal dike height can be determined by economic optimization, as first described by Van Dantzig (1956). Assume that H_0 is the current dike level and that we want to determine the amount X by which the dikes must be raised to the height H (see Figure 2).

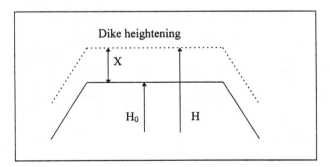

Figure 2. Schematic Overview

Let h at any moment denote the sea level along the dikes; then no loss is incurred as long as h<H. If h>H then we assume a loss with an amount of W (i.e., loss in Dutch guilders) including migration costs of the population and cattle, privation to production, damage to houses, buildings, and industry, loss of income, and other consequences.

The probability distribution of the sea level h will be denoted by F(h). The way to derive probability distributions for sea levels has been extensively examined by many authors. Good overviews are given, for example, in Russell (1982), Vrijling (1994), and Van Gelder (1996).

With a dike height of H, each year the expected loss is given by (1-F(H))W. If we assume a constant rate of interest, the expected value of all future losses is given by:

$$R = (1-F(H)) \sum_{t=0}^{\infty} (1+\delta)^{-t} = (1- F(H))W/\delta$$

The total costs of heightening a dike will be assumed as a linear function of X:

$$I = I_0 + I' X$$

where I_0 is the initial cost and I' the subsequent cost of heightening per meter. The economically optimal dike height follows by minimizing the expression R+I over the variable H (or X). In this univariate setting it is given by:

$$H_{opt} = f^{-1}(-I' \delta/W)$$

in which f^{-1} is the inverse function of the probability density f(h).

The application of the minimum total cost concept as a basis for design has gained a lot of attention lately, not only in hydraulic engineering but also in many other areas, such as structural, nuclear, and aviation engineering, among others. For example, in structural engineering good recent overviews are given by Kanda et al. (1996) and Elms (1997). The economic optimization procedure can also be used in statistical and model uncertainty, as described in Van Gelder et al. (1997).

Connecting Water-barriers

General

Some dams in the Netherlands are built to separate a sea arm or a river branch from the influence of the sea or another river. These dams, called connecting water-barriers, do not directly protect land or people from flooding, but reduce the water levels in the enclosed water body indirectly. Because of its influence on the water body behind it, the connecting water-barrier is an essential part of the system, influencing the safety of the adjacent dike ring areas.

In 1996, the Flood Protection Act was passed. This act describes the dike ring areas and sets a safety standard for each dike ring. The act defines no safety standards for the connecting water-barriers.

Immediately after the 1953 disaster, the Delta Committee was installed. Based on its recommendations (1960), the coastline of southwest Holland was made considerably safer by locking several sea-arms, thereby also reducing the length of the Dutch coast that becomes exposed directly to high storm surges. These connecting dams were supposed to be safe enough to protect the dike ring areas behind them from major disturbances of the North Sea. Hence, North Sea influences on the water systems in these areas were not taken into account.

In the 1980s the storm-surge barrier in the Eastern Scheldt was constructed. This barrier was designed to be closed only under extreme conditions. Because of this choice, the conditions behind the locked sea arm are influenced, to some extent, by the sea in all but the most extreme storm conditions.

This year, the storm-surge barrier in the New Waterway will come into use. Its design is based on the same philosophy as the design of the Eastern Scheldt barrier. However, a change can be seen from the relatively strong connecting water-barriers designed to leave the sea water outside in the past, towards a present coupled system of the connecting water-barriers with interaction between sea and lake.

The Afsluitdijk Dam

As mentioned in the introduction, the improvement of the Afsluitdijk must be completed before the year 2000. To anticipate future developments in risk-based design, the required safety level of the Afsluitdijk is the subject of a research project. In this study, the optimal safety level of the Afsluitdijk is determined by a risk-based approach. The optimization of the safety level is carried out based only on protection against flooding. Other aspects of the Afsluitdijk, such as its essential role in the traffic system of the Netherlands, are not taken into account in this study.

Economic Optimization of the Dike Height of Connecting Water-barriers

The univariate setting described before can be extended to a bivariate setting as will be shown in this section. Consider the schematic overview of the sea-lake setting of a water connecting barrier (Figure 3):

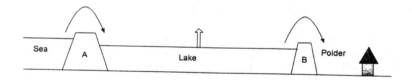

Figure 3. Schematic Overview of the Sea-Lake Setting of a
Water Connecting Barrier

We introduce the following variables:

h_A	: Height of dike A
h_B	: Height of dike B
I_{0A}	: Initial investments or mobilization costs for dike A (Dfl/km)
I_{0B}	: Initial investments or mobilization costs for dike B (Dfl/km)
L_A	: Length of dike A which has to be heightened (km)
L_B	: Length of dike B which has to be heightened (km)
dI/dh_A	: Marginal costs of dike-heightening of dike A (Dfl/km/m)
dI/dh_B	: Marginal costs of dike-heightening of dike B (Dfl/km/m)
Δh_A	: Required heightening of dike A (m)
Δh_B	: Required heightening of dike B (m)
S_A	: Damage due to a breach in dike A
S_B	: Damage due to a breach in dike B
P_A	: Probability of failure of dike A
P_B	: Probability of failure of dike B
δ	: Rate of interest

The bivariate optimization problem now becomes:

$$\min_{\Delta h_A, \Delta h_B} \{L_A I_{0A} + L_B I_{0B} + \Delta h_A L_A dI/dh_A + \Delta h_B L_B dI/dh_B + P_A S_A / \delta + P_B S_B / \delta\}$$

Failure of dike B will be more likely to happen if dike A has failed before. In that case, not only the influence of the lake but also of the sea will threaten dike B. Therefore the failure event of dike B is related to the failure event of dike A. The relationship can be described in an event tree as given in Figure 4.

<u>Description of the Hydraulic System</u>

The main hydraulic parameters which describe the loads on water-barriers are water levels and wave characteristics. They will be discussed here in the context of a sea-lake environment.

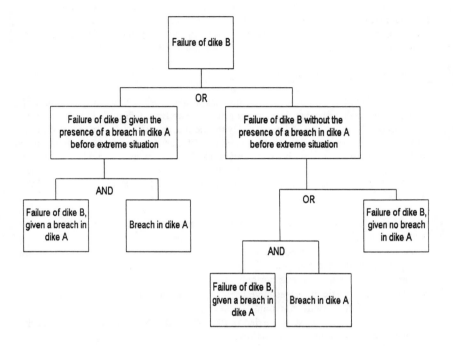

Figure 4. Event Tree of Dike B Failure

The water level along dikes around lakes consists mainly of two parameters. The first is the volume of water in the lake, influenced by river discharge, precipitation, and other factors. When a breach occurs in the connecting water-barrier (dike A in Figure 3), the volume of water in the lake is strongly influenced. The second parameter is the water level set-up (heightening) induced by high wind speeds.

The wave height along the dikes around lakes consists mainly of three parameters. The wind speed is the driving force. The fetch (the length over which the wave height is developing) and the water depth are two other important parameters. Often numerical models will be used to describe the water level and the wave height.

Several concepts can describe the strength of a dike. In the Netherlands, usually an overtopping criterion is used. The volume of overtopping depends on the water level, the wave characteristics, and the geometry of the dike. The critical volume of overtopping depends on the resistance of the dike, e.g., the grass quality.

Numerical models are available which describe the interaction between water level and wave height on the one hand and the overtopping volume on the other, but for use in probabilistic models the calculations are time-consuming. Often analytical formulae are used, based on measurements in scale models.

Methodology to Find Dominating Aspects

In general, the system with connecting water-barriers is a complex coupling of dikes and water systems. Determining the probability of failure, Pf, of the dikes along Lake IJssel is time-consuming. Hydraulic and statistical boundary conditions which are not present have to be generated. Water movement and probabilistic models have to be made. It is important to find out which variables are really necessary for determining the optimal safety level. Moreover, the technical problem is only one of the reasons argued for keeping or changing the safety standards of the Afsluitdijk.

The methodology to find the dominating aspects of the system will consist of several steps:
- A scan of the total system results in three categories of variables. These are logically dominant variables (such as wind-speed etc.), probably dominant variables, and logically unimportant variables.
- Some probabilistic exercises have been carried out to find out the most important variables of the system containing inherent uncertainty.
- A sensitivity study must be carried out to extract the dominating variables of the system containing model uncertainty.
- A guess at the uncertainty bounds of these variables must be combined with the results of the sensitivity study.
- A ranking should be made of the relative importance of all relevant variables.

Based on the results of these exercises, a decision can be made as to which subjects require research in order to get an optimal safety level which is accurate enough.

The Hydraulic System of Lake IJssel

A hydraulic model that describes the sea-lake interaction was developed by Delft Hydraulics (Den Heijer 1997).

In the case study of the Afsluitdijk, the system consists of the North Sea, Wadden Sea, the Afsluitdijk, Lake IJssel, and the dikes around Lake IJssel. The safety level of the Afsluitdijk influences the hydraulic circumstances in Lake IJssel. This can be explained by two extreme situations. First, if a stringent safety standard is chosen for the Afsluitdijk, this results in a relatively low required dike height around Lake IJssel. If, on the contrary, a less stringent safety standard is set for the

Afsluitdijk, the required dike height around Lake IJssel will have to be relatively high. See also Figure 5.

The calculation of the probability of failure for the Afsluitdijk and the dikes of Lake IJssel is treated in a system which takes into account all relevant factors. The bases for the system are the aspects of wind strength, sea-water level, tide, storage capacity of Lake IJssel, and discharge of the River IJssel. These aspects, treated as stochastic variables, interact with each other through physical relationships implemented in the water movement model.

This model couples the hydraulic system of the Wadden Sea, the Afsluitdijk dam, and the hydraulic system of Lake IJssel in a simple model for the occurrence of breaches and breach growth. The model for water movement is based on the hydraulic boundary conditions and the storage capacity of Lake IJssel. The hydraulic circumstances near the dikes around the lake are calculated by predefined reproduction functions. These functions are fits on calculations, with numerical models for several locations on Lake IJssel based on theoretical functions. With these reproduction functions, hydraulic parameters for a number of locations around Lake IJssel can be calculated as a function of global variables such as wind speed, etc.

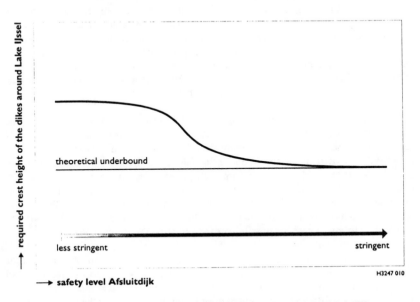

H3247 010

Figure 5. Interaction Between the Safety Level of the Afsluitdijk
and the Required Height of the Dikes around Lake IJssel

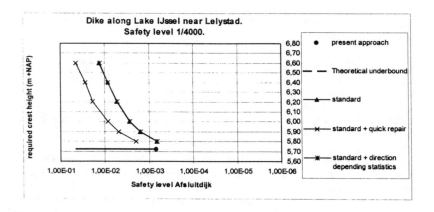

Figure 6. Influence of the Safety Level of the Afsluitdijk
on the Required Crest Height of a Dike along Lake IJssel near Lelystad

A standard calculation was carried out, with assumptions as mentioned above. The available distributions of the stochastic parameters of the model were taken into account. The influence of the safety level of the Afsluitdijk on the required crest height is visible. If the safety level of the Afsluitdijk is taken less stringently, failure of the Afsluitdijk will occur more often, and so situations 2 and 3 will also occur more often. This results in the raising of the water level on Lake IJssel, which will result in a higher required crest height of the dikes around it.

For evaluating the sensitivity of the model to some parameters, two different calculations have been carried out. The first calculation was made introducing a repair time of the Afsluitdijk of 1 month after the occurrence of a breach. This means the breach is closed in the first month after failure. In the standard calculation this repair time was variable (repair takes place in the first summer after the occurrence of a breach). The impact on the required crest height is clearly visible. This is also the case in Figure 7, in which the results of the same calculations for the second location (Enkhuizen) are presented. It is clear that in the next phase it will be worthwhile to investigate the possibility of minimizing the repair time.

The second calculation was made with wind-direction-dependent statistics for set-up of the water level on the Waddensea. In Figure 6 (calculation for Lelystad location), the model is shown to be not very sensitive for this modification. Compared to the difference in required crest height introduced by the repair time, the differences introduced by the wind-direction-dependent statistics are not relevant. On an absolute scale, a difference in required crest height of several centimeters is of the same order of magnitude as the calculating precision, and therefore not important.

Approach to Determine Optimal Safety Level of the Afsluitdijk

As described before, until now the safety standards in the Dutch Flood Protection Act have been expressed as frequencies of exceedance of water levels. These frequencies are to some extent dependent on the consequences of flooding. The dikes around Lake IJssel must be able to withstand the hydraulic circumstances with a frequency of exceedance equal to the safety standard, e.g., 1/4000 for Urkerhoek and 1/10000 for Enkhuizen.

Thus, because the probability of failure of the dikes around Lake IJssel is not a variable parameter, the minimization of the costs results in

$$\min_{\Delta hA, \Delta hB} \{ L_A I_{0A} + L_B I_{0B} + \Delta h_A L_A dI/dh_A + \Delta h_B L_B dI/dh_B + P_A S_A/\delta + P_B S_B/\delta \}$$

Under the boundary condition, in which the subscript A stands for the Afsluitdijk and B stands for the dikes around Lake IJssel, PB<1/4000 for Urkerhoek and P_B <1/10000 for Enkhuizen.

First Results and Analysis

To determine dominating aspects, a series of calculations were made of the probability of dike failure around Lake Ijssel. For this goal, the model was simplified and some assumptions were made. The most important assumptions are:

- the breach(es) in the Afsluitdijk reach a length of 1500 m. This assumption is based on experiences from the 1953 disaster, where the total length of breaches was approximately 5% of the total length of dikes in the area.
- the failure criterion of the Afsluitdijk (usually a combination of a few possible failure mechanisms, i.e., wave overtopping, geotechnical stability, and erosion) is simplified to the following definition: the Afsluitdijk fails if the water level exceeds a certain level. This level is an input parameter in the probabilistic calculations. The safety level of the Afsluitdijk equals the frequency of the exceedance of this water level.

For determining the dominating aspects, the optimization was carried out for two locations on Lake IJssel (see Figures 6 and 7).

Figure 6 visualizes the results of the calculations of the above-mentioned model for the Lelystad location (see Figure 1). Under the present safety philosophy, the Afsluitdijk is supposed to have a safety level of $7.0*10^{-4}$. The theoretical underbound for the crest height is given by the calculated probability $(1-p_{br})$ P_{F1}, which can be simplified to P_{F1}, since p_{br} is small.

The goal of the study is to determine an optimal safety level, so that the people and the land behind the dikes around Lake IJssel are as safe as required according to the Flood Protection Act. In the described system, three dangerous situations are possible:

1. The Afsluitdijk does not fail given the hydraulic circumstances. The dikes on Lake IJssel fail due to a combination of both the water level in Lake IJssel and wind-induced set-up and waves.

2. The Afsluitdijk fails given the hydraulic circumstances at the Wadden Sea. A breach is being formed, and a certain quantity of water flows into Lake IJssel. The storm surge is transmitted into the lake and threatens the dikes on Lake IJssel.

3. The Afsluitdijk has failed earlier, and a breach has formed. On Lake IJssel delayed tidal movement takes place with a relatively high average water level. This situation can last for a certain period depending on the repair time. Because of the raised level of Lake IJssel, a small storm can be a serious threat to the safety of the dikes all around Lake IJssel.

Given the hydraulic load on the Afsluitdijk, one out of three different dangerous hydraulic situations can occur on Lake IJssel.

The total probability of failure of a dike around Lake IJssel is given by:

$$P_B = P_{B1} (1-P_A)+P_A (P_{B2} +T_A (1-P_{B2})P_{B3})$$

in which:

P_B = Probability of failure of a dike along Lake IJssel.
P_{B1} = Probability of failure of a dike along Lake IJssel in situation 1.
P_{B3} = Probability of failure of a dike along Lake IJssel in situation 2.
P_{B3} = Probability of failure of a dike along Lake IJssel in situation 3.
P_A = Probability of failure of the Afsluitdijk.
T_A = Factor representative of the repair time. When repair is ready in the first summer after a breach, T_r = 1/2; when repair is ready within one month after a breach, T_r = 1/6.

The probabilities P_{B1}, P_{B2}, and P_{B3} are calculated with a First-Order Reliability Method (FORM), as described in, for example, Ang and Tang (1990). A model based on numerical integration called HYDRA-M (1996) built by the Ministry of Water Management to consider the first situation, was used to verify the FORM calculations, with promising results.

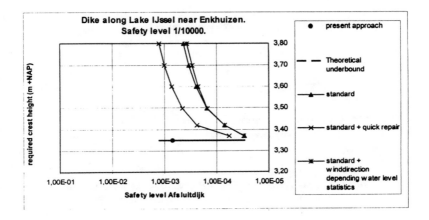

Figure 7. Influence of the Afsluitdijk on the Required Crest Height of a Dike along Lake IJssel near Enkhuizen

The differences between the standard and the direction-dependent calculations, shown in Figure 7, are even smaller. This is very surprising because maximal water level set-up and maximal wind speeds are normally from different wind directions. The derivation of direction-dependent statistics seems to get low priority in the next phase of the research project. However, in the future one more location will be studied before this conclusion may be drawn.

Some results are displayed in Table 1. The required crest height above the theoretical underbound is presented for the two locations. This underbound is the safety of the dike for a specific location in situation 1, as described earlier.

Table 1. The Difference Between the Required Crest Height and the Theoretical Underbound

Safety level Afsluitdijk Location	1/1430	1/4000	1/10000
Enkhuizen	0.40	0.15	0.05
Lelystad	0.10	0.05	-

In the next part of the study other important parameters will be varied, such as the breach dimensions.

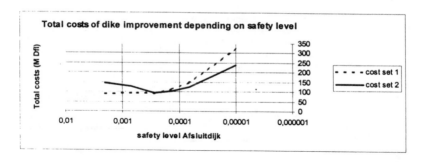

Figure 8. Two Variants of the Total Costs of Dike Improvement as a Function of the Safety Level of the Afsluitdijk

In Figure 8, the optimization of the safety level of the Afsluitdijk has been carried out. For several safety levels, the required height improvement of the dikes around Lake IJssel can be read from Figures 6 and 7. For the optimization, some very coarse estimates of the cost of heightening have been made. Also, the lengths of the dikes needing improvement are estimated. Therefore, the results of the optimization are tentative. Two different estimated cost sets were used. In the first, the improvement of the Afsluitdijk is twice as expensive as the improvement of a dike along Lake IJssel, while in the second set these costs are equal. Also, the total length of dikes along Lake IJssel which will need to be improved cannot yet be estimated correctly. For certain levels of safety for the Afsluitdijk, the height improvement of dikes along Lake IJssel has been deduced from Figures 6 and 7. If the safety level of the Afsluitdijk is to be higher than at present, improvement is necessary. Clearly the present situation of both the Afsluitdijk and of the dikes around Lake IJssel are very important for the optimization. Some dikes around Lake IJssel are former sea dikes, designed to resist storm surges, and therefore higher than strictly required. Clearly visible is the different minimal cost level for the different cases. Furthermore, the optimal safety level in case 1 is different from the optimum derived from case 2. In the figure we see that for safety levels below 1/10000 the total costs don't vary significantly compared to the costs for more stringent safety levels.

Conclusions and Recommendations

In this paper, a systems approach has been described to determine the optimal safety level of connecting water-barriers in a sea-lake environment. It is based on a bivariate economic minimization of the expected total costs. A case study of the Afsluitdijk in the Netherlands has been described.

Calculating the probability of failure of a dike on Lake IJssel, influenced by both the hydraulic conditions on the lake and the safety level of the Afsluitdijk dam, is possible with rather simple tools. Implementation in a First-Order Reliability Method (FORM) gave rather promising results. The calculations are accurate and fast. Some dominant factors can be determined after carrying out the calculations for two locations. A sensitivity study can determine the other dominant factors.

The direction-dependent statistics for water level set-up seems not to be an important factor compared with use of the wind-direction-dependent statistics. The repair time, on the contrary, has a significant influence on the results. This parameter should be explored in the next phase of the research project. In the next part of this phase of the study other parameters will be varied, such as the breach dimensions.

A simple exercise to get some insight into the important parameters in the optimization of the safety level of a connecting water-barrier, based on minimizing the investments, leads to the following conclusions:

- It is necessary to have a good knowledge of the parameters determining the cost of a dike improvement.
- The length of the dikes which will need improvement is an important quantity.
- Contrary to the approach described in this paper, all individual sections of the dikes along Lake IJssel and the Afsluitdijk dam should be taken into account in a thorough risk-based optimization of the safety level of the Afsluitdijk.

The method described in this paper provides a risk-based optimization of the safety level of a connecting water-barrier. The approach is based on technical and statistical grounds. For an integral determination of the safety level, other factors such as the traffic function of the Afsluitdijk must be taken into account.

Acknowledgements

The authors acknowledge the comments of Professor Hans Vrijling from Delft University of Technology, Geoff Toms from Delft Hydraulics, and Harry de Looff from the Ministry of Water Management, Public Works and Transport.

References

Ang, A.H.S., and W.H. Tang. 1990. *Probability Concepts in Engineering, Planning and Design, Vol. II (6)*. New York: John Wiley and Sons.

Delta Committee. 1960. *Report of the Delta Committee*. The Hague, The Netherlands.

Den Heijer, F. 1997. *Required Safety Level of the Afsluitdijk.* Technical report, Delft Hydraulics (forthcoming).

Elms, D.G. 1997. Risk balancing in structural problems. *Structural Safety* 19(1): 67-77.

HYDRA-M.1996. Computer program for numerical integration. Ministry of Water Management, Public Works and Transport, The Netherlands.

Kanda, J., and K.A. Ahmed. 1996. Optimum reliability-based design earthquake load. *Probabilistic Mechanics and Structural and Geotechnical Reliability.* Proceedings of the Specialty Conference 1996:194-197. New York: ASCE.

Russell, S.O. 1982. Flood probability estimation. *Journal of the Hydraulics Division* 108(HY1), January.

Van Agthoven, A.M., F. Den Heijer, and A.W. Kraak. 1997. *The Way to a Flood Risk Safety Concept: Four Case Studies.* Delft, The Netherlands: European Union.

Van Dantzig, D. 1956. Economic decision problems for flood prevention. *Econometrica* 24: 276-287, New Haven, CT.

Van Gelder, P.H.A.J.M. 1996. A new statistical model for extreme water levels along the Dutch coast. *Stochastic Hydraulics* 243-250, Mackay, Australia: A.A. Balkema.

Van Gelder, P.H.A.J.M., J.K. Vrijling, and K.A.H. Slijkbuis. 1997. Coping with uncertainty in the economic optimization of a dike design. *Proceedings of the 27th IAHR Congress: Water for a Changing Global Community,* San Francisco, CA.

Vrijling, J.K. 1994. Sea-level rise: a potential threat? *Statistics for the Environment 2: Water-Related Issues.* New York: John Wiley and Sons.

Risk-Averse Reliability-Based Optimization of Sea Defenses

P.H.A.J.M. van Gelder[1] and J.K. Vrijling [2]

Abstract

Engineers' design preferences may change with the amount of monetary value that is involved in a structural system. When there could be a large amount of damage due to the possible collapse of a structure, the engineer will be very risk-averse in his design. In this paper it will be shown how risk-averse behavior can be included in a reliability-based design by using convex utility functions of monetary value. The procedure will be applied to a design of a sea defense system.

Introduction

The reliability concept is now widely accepted in civil engineering as a tool to find the most appropriate structural system in terms of safety and economy. Design engineers can determine the optimum degree of safety for a structure to satisfy the safety demands and economic concerns of owners or users. Reliability-based optimal structural design in the field of water resources has been applied by numerous authors, e.g., Van Dantzig (1956), Bernier (1987), and Stedinger (1997). In the field of structural engineering much work has been done by Grigoriu et al. (1979), Frangopol (1985), and Kanda et al. (1997).

One of the difficulties in applying a reliability-based optimization lies in the lack of supporting data that can be used in the associated probability-based decision models. Also, the uncertainty in the economic parameters of structural systems can cause difficulties in optimization procedures. Both problems have been approached in the recent work of Slijkhuis et al. (1997) and Van Gelder et al. (1997a). A

---

[1]PhD Candidate, Delft University of Technology, Faculty of Civil Engineering, Stevinweg 1, 2628 CN Delft, The Netherlands
[2]Professor, Delft University of Technology, Faculty of Civil Engineering, Stevinweg 1, 2628 CN Delft, The Netherlands

proposal to deal with time-variant probabilities of failure in a reliability-based optimization was suggested by Vrijling et al. (1997). In Ang and Tang (1990) it was already noted that design preferences of engineers may change with the amount of monetary value that is involved in the structure. When the amount of damage due to a possible structural collapse is huge, the engineer will be very risk-averse. In this paper, the concept of risk aversion will be included in the reliability-based design. The use of convex utility functions of monetary value will be shown to be appropriate in modeling the behavior of the risk-averse engineer. The procedure will be applied to a design of a vertical breakwater. Vertical breakwaters have become quite popular as a sea defense system for large harbors. Designing vertical breakwaters by a reliability-based optimization was shown to be successful by Voortman (1997). The amount of structural and economic damage due to the collapse of a vertical breakwater is enormous. Including risk-aversion in the structural design is therefore very important.

The paper is organized as follows. First, we explain the function and failure modes of vertical breakwaters and the concept of reliability-based optimal structural design. In a reliability-based design, it is essential to describe the consequences of the failure of vertical breakwaters, and they will therefore be analyzed in detail for a fictitious example. The construction costs of vertical breakwaters also play an important role in the procedure, and they are analyzed carefully in the paper. One failure mechanism of vertical breakwaters, namely overtopping, will have our particular attention. The concept of risk aversion is explained and applied to the fictitious example. Finally, conclusions are drawn.

<u>Vertical Breakwaters</u>

A vertical breakwater is used to protect a harbor basin against wave action; i.e., its main function is to provide a tranquil harbor basin. Figure 1 shows a vertical breakwater at the harbor of Scheveningen in the Netherlands.

Figure 1. Vertical Breakwater in Action (Photograph by W. Kuiper)

The failure mechanisms of a vertical breakwater are of two types (De Groot et al. 1996):
• final failure modes and
• preceding-failure modes.

Final failure modes are those which lead directly to the top event (for instance, severe overtopping). Preceding-failure modes do not lead directly to the top event, but bring the structure into a dangerous state or induce a final failure mode (for instance, erosion of the seabed in front of the rubble foundation.) Failure conditions are not failure modes, but induce failure modes to occur (for instance, high wave load).

The *final failure modes* (Figure 2) cause failure of the structure These are:
• sliding of the structure over the foundation,
• bearing capacity failure of the foundation,
• disintegration of (a part of) the structure,
• settlement due to densification of the foundation soil or internal erosion, and
• overtopping or wave transmission.

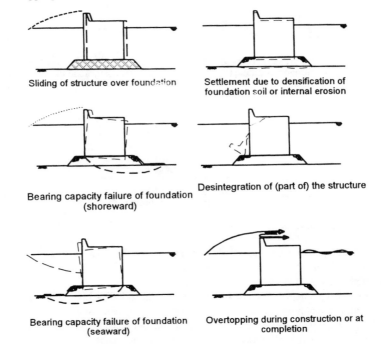

Figure 2. Overview of Final Failure Modes (Taken from De Groot et al. 1996)

The *preceding-failure modes* bring the structure into a dangerous state or (eventually) cause the occurrence of a final failure mode (see Figure 3). The preceding-failure modes are:

- erosion of the rubble foundation,
- erosion of the seabed at the toe or rubble foundation,
- internal erosion of rubble foundation or subsoil, and
- loss of caisson fill after disintegration of caisson.

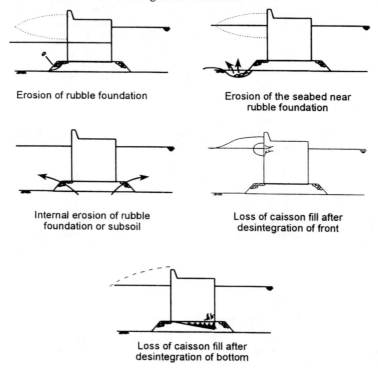

Erosion of rubble foundation

Erosion of the seabed near rubble foundation

Internal erosion of rubble foundation or subsoil

Loss of caisson fill after desintegration of front

Loss of caisson fill after desintegration of bottom

Figure 3. Overview of Preceding-Failure Modes (De Groot et al. 1996)

Failure conditions are loads or load effects. These cause final or preceding-failure modes to occur. The failure conditions are:

- high wave load to wall,
- high instantaneous pore pressure in foundation, including high uplift force,
- high residual pore pressures and/or degradation in the subsoil,
- high wave pressures along rubble foundation and seabed, and
- high current velocities along rubble foundation and seabed.

A fault tree for the breakwater down to the level of the final failure modes is given in Figure 4. The failure mode which is used in the optimization calculations of this paper is the SLS (serviceability-limit-state) wave transmission. Its reliability function is given by

$$R = WL - h_c$$

in which WL is the random variable of water levels in front of the vertical breakwater, to be determined from a historical record and/or physical models (see Van Gelder et al. 1997b).

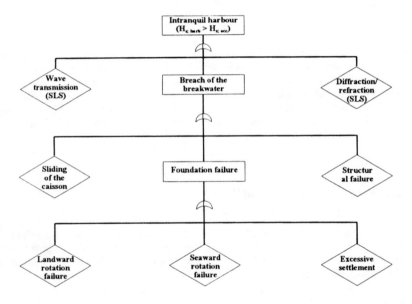

Figure 4. Fault Tree of a Vertical Breakwater

Reliability-Based Optimal Structural Design

Heightening of a structure

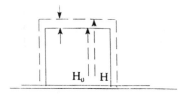

Figure 5. Schematic Overview

The principal of a reliability-based optimal structural design can be explained with the following illustrative example (see Figure 5). Assume that an existing structure has a height of H_0. The structure will be heightened to an optimal height H. There are costs involved with this heightening, which are a function of X where $X = H - H_0$. These costs can be assumed dependent on X by the relation $I = I_0 + f(X)$, in which I_0 are the mobilization costs and f(X) are the costs-per-meter dike heightening. The function I increases with the height. There are also costs involved due to a possible failure of the structure. These costs are given by $P(failure)S/\delta$ in which S is the cost of damage due to failure, δ is the rate of interest-growth and P(failure) the probability of failure. Note that the P(failure) function decreases with the height. Summation of both types of costs gives the objective function which must be minimized over the design variable H in order to find the optimal design:

$$\text{Min}_{\Delta h} \{ I_0 + \Delta h \; dI/dh \; + P(failure)S/\delta \}$$

Construction Costs of Vertical Breakwaters

To optimize a breakwater design, the total lifetime costs are written as a function of the design variables. In this study the design variable used is the height of the caisson. The input for the cost function consists of estimates of the construction costs and the costs in case of failure. As stated before, the lifetime costs consist of two main components:

- the construction costs, and
- the risk.

Part of the construction costs are due to the breakwater geometry (variable costs), and part can only be allocated to the project as a whole (project costs). For a breakwater caisson, the variable costs can be assumed to be proportional to the volumes of concrete and filling sand in its cross section. Therefore, the construction costs can be written as (Voortman 1997):

$$I_{constr}(h_c, B_c) = I_0 + I_{sand} V_{sand}(h_c, B_c) + I_{concr} V_{concr}(h_c, B_c)$$

in which:
h_c = height of the caisson [m]
B_c = width of the caisson [m]
I_0 = fixed costs [units]
I_{sand} = costs of filling sand [units/m^3]
V_{sand} = volume of filling sand in the breakwater [m^3]
I_{concr} = costs of concrete [units/m^3]
V_{concr} = volume of concrete in the breakwater [m^3]

The yearly risk is defined as the expected value of the damage costs per year. In formula:

$$R_{year}(h_c, B_c) = 365\,P_{f;SLS}(h_c, B_c)\,C_{SLS} + P_{f;ULS}(h_c, B_c)\,C_{ULS}(h_c, B_c)$$

in which:

$P_{f;SLS}(h_c, B_c)$ = Probability of serviceability-limit-state (SLS) failure *per day* as a function of the width and height of the caisson [1/day]
C_{SLS} = Costs per day in case of serviceability-limit-state failure [units/day]
$P_{f;ULS}(h_c, B_c)$ = Probability of ultimate-limit-state (ULS) failure *per year* as a function of the width and height of the caisson [1/year]
$C_{ULS}(h_c, B_c)$ = Costs per event in case of ultimate-limit-state failure as a function of the width and height of the caisson [units/event]

The costs in case of ULS failure consist of replacement of (parts of) the breakwater and thus depend on the caisson dimensions. The costs in case of SLS failure are determined by the costs of downtime and thus are independent of the caisson geometry.

The total risk over the lifetime of the structure is given by the sum of all yearly risks, corrected for interest, inflation, and economic growth. This procedure is known as capitalization. In formula:

$$R_{lifetime}(h_c, B_c) = \sum_{n=1}^{N} \frac{R_{year}(h_c, B_c)}{(1 + r' - g)^n}$$

in which:
r' = net interest rate [%]
g = growth rate [%]

The growth rate expresses that in general the value of all goods and equipment behind the breakwater will increase during the lifetime of the structure.

The function which describes the lifetime costs can finally be written as:

$$C_{lifetime}(h_c, B_c) = I_{constr}(h_c, B_c) + R_{lifetime}(h_c, B_c)$$

Minimization of this function simultaneously results in the optimal probability of failure and the optimal caisson dimensions.

When observing a breakwater cross-section, several cost components can be discerned (Figure 6). The cross-section consists of two main components: the caisson, and the rubble foundation. In this study, only the height of the caisson is used as a design variable; therefore, the costs of the rubble foundation are omitted. The costs of the caissons are in principle determined by:
- the volume of concrete in the caisson cross section: V_{concr};
- the volume of filling sand in the caisson cross section: V_{sand};
- the total length of the breakwater: L_{brwt}.

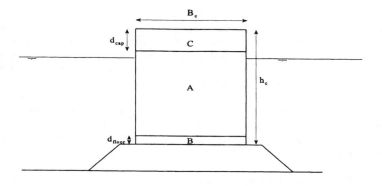

Figure 6. Parts of the Caisson Cross Section

Usually, a breakwater caisson is equipped with dividing walls. The number of walls depends on the width of the caisson. When optimising the breakwater design for the width of the caisson, the number of dividing walls is unknown. Therefore, in principle, the volumes of concrete and filling sand are unknown. This problem can be solved by assuming a fixed percentage of concrete in the caisson cross-section in area A (see Figure 6). The thickness (and thus the area) of the floor and concrete cap (areas B and C) are set to fixed values. The volumes of concrete and filling sand now can be calculated in the following way:

$$V_{concr} = B_cL_{brwt}[\{(d_{cap} + d_{floor}) + p(h_c - d_{cap} - d_{floor})\}]$$
$$V_{sand} = h_cB_cL_{brwt} - V_{concr}$$

in which p denotes a fixed percentage of concrete.

The costs of concrete and sand are as given in Table 1.

Table 1. Overview of Material Costs

Material	Price [US $/m^3]
Filling sand	5
Concrete	250

Several cost components cannot be allocated to the breakwater cross section, but only to the building project as a whole. Examples of these cost components are:
- costs of the feasibility study,
- costs of the design of the breakwater,
- site investigations, such as penetration tests, borings and surveying, and
- administration.

These costs are included in the cost function by means of a fixed sum of money per metre breakwater (I_0).

In principle, there are two ways in which a breakwater can fail. Either it collapses under survival conditions after which there will be more wave penetration in the protected area, or the breakwater is too low and allows too much wave generation in the protected area due to overtopping waves. In both cases, the harbor operations in the area of interest possibly will have to be stopped, resulting in damage (downtime costs).

The exact amount of downtime costs is very difficult to determine. The downtime costs for one single ship can be found in the literature (see, for instance, DUT 1995), but the total damage in case of downtime does not depend solely on the downtime costs of ships. The size of the harbor and the type of cargo are also important variables in this type of damage. Furthermore, the availability of an alternative harbor is very important. If there is an alternative, ships will make use of it. In that case, the damaged harbor will lose income as fewer ships use it and possibly also because of shipping company claims. On a macro-economic scale however, there may be minor damage since the goods are still coming in by way of the alternative harbor. Thus, the availability of infrastructure in the area influences the damage in case of downtime. If an alternative harbor is not available, the economic damage may be felt beyond the port itself.

The location of the breakwater in relation to the harbor also influences the damage costs. If the breakwater protects the entrance channel, the harbor cannot be reached during severe storms, thus causing waiting times. These waiting times can range from hours to a few days. If the breakwater protects the harbor basin, terminal damage to the breakwater can cause considerable amounts of extra downtime due to the fact that the structure will only partly fulfill its task over a longer period of time. If the load on a structural component exceeds the admissible load, the component will collapse. Several scenarios are now possible:

1. The component is not essential to the functionality of the breakwater. Repair is not carried out and there is no damage in monetary terms. This is the case if, for instance, an armor block is displaced in the rubble foundation. It should be noted that this kind of damage can cause failure if many armor blocks are displaced (preceding-failure mode).
2. The component is essential to the functioning of the breakwater, but the stability of the caissons is not threatened. This is the case if, for instance, the crown wall collapses. The result is a reduction of the crest height of the breakwater, which could threaten its functioning. Therefore, repair must be carried out and there is some damage in monetary terms.
3. The caisson has become unstable during storm conditions. There is considerable damage, resulting in necessary replacement of (parts of) the breakwater. The damage in monetary terms is possibly even higher than the initial investment in the breakwater.

The examples in Tables 2 and 3 have been taken from Vrijling (1997). The costs are considered from the point of view of the Port Trust (PT) harbor authorities.

Table 2. Fictitious Example: Costs of One-Day Suspension of Harbor Operations

Item	Description	Cost [US $]
Costs of shipping operation	US $10,000 per vessel per day; average 1 vessel per day; waiting 3 days extra	30,000
Loss of income PT, direct	Throughput 19 mln t/yr; port dues US $5.4/t	280,000
Loss of income PT, indirect	Bad reputation per day	140,000
Claims	Industry, shipping lines, others	50,000
	SUBTOTAL	500,000
Multiplier for indirect economic damage	1.5	250,000
	TOTAL	750,000

Table 3. Fictitious Example: Total Costs in Case of Failure of the Breakwater

Item	Description	US $million
Structural Damage		
Damage to the breakwater	20% of construction costs	24
Damage to other structures in the harbor	Wharf, slope protection, harbor lights	5
Mobilization	Lump sum	4
	SUBTOTAL	33
Economic Damage		
Alternative transportation of inputs to industry	Throughput 19 mln t/yr; extra transport costs per ton: US $6	114
Cost of shipping operation	US $10,000 per day; average 1 vessel per day	3.65
Loss of income PT, direct	Throughput 19 mln t/yr; lost port dues US $5.4/t	103
Loss of income PT, indirect	Bad reputation	50
Lives lost	< 10, economic damage negligible	-
Claims	Industry, shipping, other parties	100
	SUBTOTAL	370
Multiplier for indirect economic damage	1.5	555
Total structural and economic damage		588

When optimising a breakwater design, an estimate of the damage is needed. In the case of a structural component, this could be the cost of rebuilding. If large parts of the caissons are collapsed, the area will have to be cleared before rebuilding the breakwater. In that case, the damage will be higher than in the case of rebuilding alone. Furthermore, in general, collapse will lead to downtime costs, which further increases the damage.

Utility Function of Monetary Value

Monetary value is not always a consistent measure of utility. Consider the following example. A decision maker is asked to choose between alternatives I and II from the follo.ving pair of lotteries (see Figure 7):

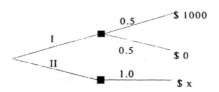

Figure 7. Pair of Lotteries

In lottery I, the outcome will be either $1000 or $0 with equal likelihoods. In lottery II, the only outcome is a sure x = $500. The expected monetary values are the same for the two lotteries. However, a decision maker might prefer lottery II, since there is a sure gain of $500, whereas in lottery I there is a 50% chance of gaining nothing. The lottery differences may not matter, depending on the financial status of the decision maker relative to the monetary values of the alternatives. Therefore, a utility function for money should be established. If the decision maker is a large firm or a government organization, the utility function over a range of monetary values may be a straight line. If the decision maker is risk-averse, the utility function will show convex behavior over the range of monetary values.

$$u(1000) = 1$$
$$u(0) \quad = 0$$

Suppose the decision maker chooses for x = 100, then

$$u(100) = 0.5\ u(1000) + 0.5\ u(0) = 0.5$$

This procedure can be continued for other monetary values in order to obtain the monetary utility curve. A series of risk-averse utility functions is given by:

$$u(x) = \frac{1}{1-e^{-\gamma}}(1-e^{-\gamma x})$$

As γ increases, the utility function becomes more convex, indicating higher risk-aversion. In Figure 8, the risk-averse utility functions for $\gamma = 1, 2$, and 3 are depicted.

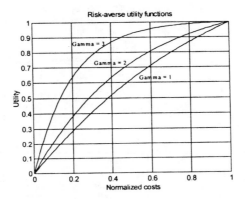

Figure 8. Convex Utility Functions

Including the concept of risk-aversion in the reliability-based optimization can be done in the following way. There is a continuous decision variable, namely h (= height of the breakwater). If a certain h is chosen, there are two possible consequences on the SLS wave transmission (R = WL - h):

- With probability $1-F_{WL}(h)$ there is a loss of $S + I_0 + I_1h$

- With probability $F_{WL}(h)$ there is a loss of $I_0 + I_1h$

The losses can be transformed with the risk-averse function for money as given above. One ends up with the following utility function which can be optimized over the decision variable h:

$$u(h) = -(1-F_{WL}(h)) \exp(-\gamma(S + I_0 + I_1h)) - F_{WL}(h) \exp(-\gamma(I_0 + I_1h))$$

Applying this approach to the fictitious example, the following results are obtained (Figure 9). Without risk-aversion (i.e. risk-neutral) the reliability-based optimal design is given by h = 4m. Applying risk-aversion with different levels (low, medium, and high; corresponding to γ = 1, 2, and 5), an increase in the optimal design height is found from 4.3, 4.4, to 4.8m.

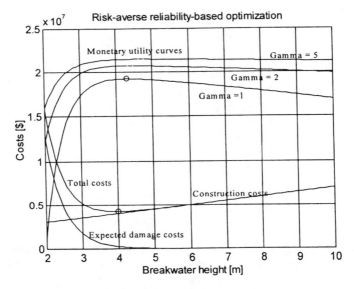

Figure 9. Results of the Fictitious Example

Conclusions

In this paper, we have described the reliability-based optimal structural design of vertical breakwaters with respect to the failure mechanism of overtopping. In preceding papers, such as Van Gelder et al. (1997a) and Slijkhuis et al. (1997), it was shown that a reliability-based design should take into account statistical and probabilistic model uncertainties. A Bayesian framework was developed for that purpose. In this paper, the reliability-based design is included with the concept of risk aversion by using utility functions of monetary value.

A fictitious example of the reliability-based optimal design of a vertical breakwater has been presented. A detailed analysis of the construction and damage costs has been given. Different levels of risk aversion have been applied in the optimal design. A higher risk aversion leads to a more conservative structural design. The exact level of risk aversion, however, is difficult to determine. It depends not only on the economic status of the decision maker, as was illustrated in this paper, but also on political and ethical issues. When human lives are involved with the failures of structures, the techniques in this paper are very difficult to apply. Recent research in the field of acceptable risks (e.g., Vrijling et al. 1996) is promising for these cases.

References

Ang, A.H.S., and W.H. Tang. 1990. *Probability Concepts in Engineering, Planning and Design--Volume II: Decision, Risk, and Reliability.* New York: John Wiley and Sons, Inc.

Bernier, J. 1987. Elements of Bayesian analysis of uncertainty in hydrological reliability and risk models. In L. Duckstein and E.J. Plate (Eds.), *Engineering Reliability and Risk in Water Resources.* NATO ASI series 124: 405-423.

De Groot, M.B., H.J. Luger, and H.G. Voortman. 1996. *Description of Failure Modes.* Technical Report, MAST-III/PROVERBS. Delft Geotechnics, Delft University of Technology, Delft, The Netherlands.

DUT. 1995. Risk analysis and probabilistic modeling. Coal project, India. Delft University of Technology, Delft, The Netherlands.

Frangopol, D.M. 1985. Structural optimization using reliability concepts. *Journal of Structural Engineering*, ASCE 111: 2288-2301.

Grigoriu, M., A.M. Veneziano, and C.A. Cornell. 1979. Probabilistic modeling as decision making. *Journal of the Engineering Mechanics Division* EM4: 585-598.

Kanda, J., and K.A. Ahmed. 1997. Optimum reliability-based design loads due to natural hazards. *Structural Engineering International* 2/97: 95-100.

Slijkhuis, K.A.H., P.H.A.J.M van Gelder, and J.K. Vrijling. 1997. Reliability-based optimization of flood protection under statistical, damage, and construction uncertainty. ICOSSAR '97, 7th International Conference on Structural Safety and Reliability, Kyoto, Japan.

Stedinger, J.R. 1997. Expected probability and annual damage estimators. *Journal of Water Resources, Planning and Management* March/April: 125-135.

Van Dantzig, D. 1956. Economic decision problems for flood prevention. *Econometrica* 24: 276-287.

Van Gelder, P.H.A.J.M., J.K. Vrijling, and K.A.H. Slijkhuis. 1997a. Coping with uncertainty in the economic optimization of a dike design. 27th IAHR Congress 1997, Water for a Changing Global Community. San Francisco, CA.

Van Gelder, P.H.A.J.M., J.K. Vrijling, and H. Voortman. 1997b. Probabilistic analysis of wave transmission due to overtopping of vertical breakwaters. *Probabilistic Tools for Vertical Breakwater Design*. MAS3-CT95-0041, Proceedings Workshop, Las Palmas, February 1997.

Voortman, H.G. 1997. Economic optimal design of vertical breakwaters. MS thesis, Delft University of Technology, Delft, The Netherlands.

Vrijling, J.K., W. van Hengel, and R.J. Houben. 1996. Acceptable risk: a normative evaluation. *Stochastic Hydraulics*: 87-94. Rotterdam, The Netherlands: Balkema.

Vrijling, J.K., and P.H.A.J.M. van Gelder. 1997. The effect of inherent uncertainty in time and space on the optimal reliability of vertical breakwaters. *Proceedings of the ESREDA Seminar, Paris, France*. To be published.

On the Application to Risk and Decision Analysis of TRIZ, the Russian Theory of Inventive Problem-Solving

Stan Kaplan, Ph.D[1]

Abstract

Traditional decision theory deals with the problem of selecting the best option out of a given predetermined set of options. It has nothing to say about where those options come from. That is the "creative" part of the problem. The role of risk analysis in decision making is to predict – i.e., to calculate probabilistically – the outcomes of each decision option. As part of that process, risk analysis must gather all the information and evidence relevant to those outcomes. Risk analysis may thus be considered the "information-gathering" option, or the "inner loop" of a decision analysis. The Russian theory TRIZ can be thought of as an "option-generating" or "outer loop" of a decision process. In this sense, TRIZ completes the decision theory model in an important way. We note that just one good idea in the outer loop of a decision problem can make unnecessary a great deal of tedious and expensive work in the inner loop. The Yucca Mountain nuclear waste repository is cited as an example.

Introduction

This paper presents a reformulation of the classic model (North 1960) of a decision problem into a form which makes clear the roles, in decision analysis, of Quantitative Risk Analysis (QRA), Bayes' Theorem, and what we call the "evidence-based" approach to risk and decision. Within this formulation, the risk-assessment process can be seen as the "inner loop" formed by the classic concept of the "information-gathering" option. The idea of an "outer loop" is introduced. This generates the options in the first place and refines them, or generates new options in response to what has been learned during the risk analysis. To show that the process

---

[1]Bayesian Systems, Inc., 6000 Executive Blvd., Suite 600, Rockville, MD 20852

of generating options can be systematized, we cite the Russian theory TRIZ as an example. A little bit of work at the outer loop level can often do away with large amounts of struggle at the inner loop level.

The outer loop, together with QRA, Bayes', and the new formulation constitutes a more complete, explicit, and satisfying model of the decision process.

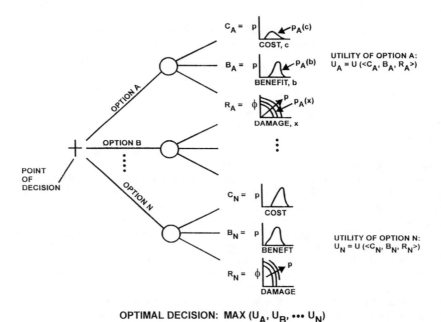

OPTIMAL DECISION: MAX (U_A, U_B, ••• U_N)

Figure 1. Anatomy of a Decision: Alternate Form

Decision Theory

The anatomy of a typical decision problem is shown in Figure 1. In this figure we represent the totality of the outcomes, of any given option, as having three components: costs, benefits, and risks. At the point of decision, if we choose option 1 for example, we do not know for sure what the cost outcome will be, since it is, after all, in the future. However, we do have some knowledge about it. As shown, we express this knowledge in the form of a probability curve over a cost axis, C. Similarly, we express our state of knowledge about the benefits to be received from option 1 in the curve against B. By *risks* we mean the likelihood of accidents, discrete events that might occur as a result of choosing this option. In the traditional format, this likelihood is expressed as a band of complementary cumulative distribution functions (CCDFs). Each CCDF shows the *frequency of exceedance* of a given level of damage, x, that is the frequency of events causing damage at level x or above. Since this frequency is also not known for certain, the *band*, or family of curves, with probability as the parameter of the family, expresses our state of knowledge about the true CCDF.

Thus, the risks are also shown by what is in effect just another form of probability curve, and the outcome of any decision option as a triplet of probability curves, <C,B,R>. Our degree of preference for the outcome is then expressed by assigning a *utility function*, U, to each triplet. The decision is then made by choosing that option whose triplet has the largest utility.

Risk Assessment, Bayes' Theorem, and the "Evidence-Based" Approach

Now, where do these triplets of probability curves come from? In Figure 2, we view the production of these curves as the job of QRA, *quantitative risk assessment.* QRA in turn obtains these curves by identifying risk scenarios and calculating the frequencies and consequences for each (Kaplan 1991). Our state of knowledge of these frequencies is also expressed by probability curves. These curves are determined by the available evidence, by processing each evidence item through Bayes' Theorem, which is the fundamental logical principle governing the process of making inferences from evidence. Thus, as we show in Figure 2, the <C,B,R> triplets are determined ultimately by the evidence – the totality of evidence available.

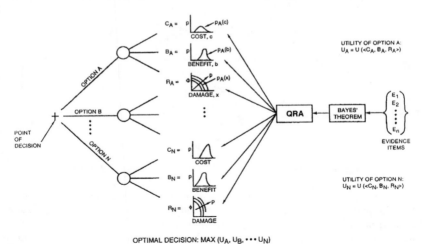

OPTIMAL DECISION: MAX (U_A, U_B, $\cdots U_N$)

Figure 2. Anatomy of a Decision: Role of Bayes' Theorem

When the probability curves are determined by the evidence in this way we call the QRA "evidence-based." A decision made on such a QRA is called an "evidence-based decision" and the evidence-based philosophy (Kaplan 1997) is summarized in the slogan: "LET THE EVIDENCE SPEAK," meaning not the personalities, opinions, positions, politics, moods, entrenched beliefs, or wishful thinking.

The Information-Gathering Option.

Suppose that the probability curves coming from the QRA are so wide that we are not comfortable choosing any of the decision options. In every decision problem there is always another option implicitly present: the *information gathering* (IG) option (North 1960). This option says, "Don't decide now, get some more information, run some more tests, etc., and then decide."

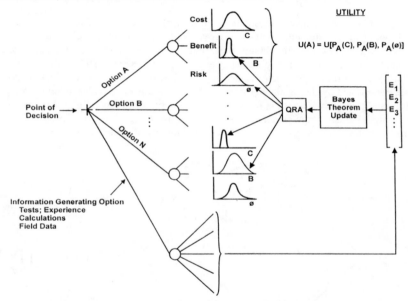

Figure 3. Option for Gathering or Generating Better Information

The IG option is shown in Figure 3. As with any other option, it has its costs, benefits, and risks. Information costs money and time, and the information produced may be untrustworthy, wrong, or simply not useful. Usually, however, it produces new evidence, which adds to the evidence base as shown in Figure 3, and which changes, and hopefully narrows, the probability curves. We are then back at the point of decision, but with new curves. The IG option may thus be thought of as an "inner loop" in the decision problem.

The Outer Loop

Classical decision theory is concerned with selecting the best out of a given set of decision options. It has nothing to say about where these options came from. This is the "creative" or "inventive" part of the problem. In Figure 4, the creation of new options is shown as what we may think of as an "outer loop." The "option generator" box represents some creative process such as the well-known "brainstorming" technique, or the more systematic Russian theory TRIZ (Kaplan 1996).

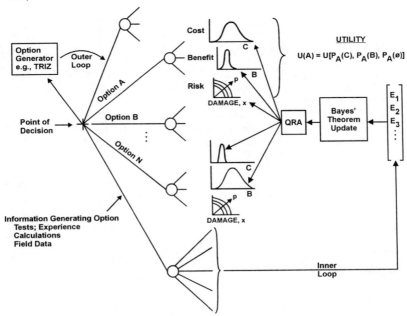

Figure 4. The "Outer Loop" of a Decision Problem: Generating New Options

The outer loop as an alternative to the inner loop

The outer loop is an alternative to the inner loop in the sense that if we are not satisfied with the existing options, we can work harder to gather more information and do a better QRA, or we can try to find a new, more satisfactory option. Thus, a little bit of creative thinking outside the bounds of our previous assumptions, can replace huge amounts of work and struggle within those bounds.

An Example

An example in point is the proposed Yucca Mountain nuclear waste repository. Huge amounts of time and talent have been and continue to be poured into QRA (there called PA, for "performance analysis") and information-gathering (there called "site characterization"), yet no decision is emerging. It is becoming clear that no amount of additional site study – geological, hydrological, vulcanological, etc. – is going to convince the State of Nevada, and others, that the geology will contain the waste with the desired degree of confidence, for the desired period of time. What is needed now (Garrick and Kaplan 1998) is to shift thinking to the outside loop, and reconsider a) containment by engineered barriers, b) isotope separation and transmutation, and even c) the entire basic mind-set of "permanent disposal."

References

Garrick, B. J., and S. Kaplan. 1998. Making decisions about the management of high-level radioactive waste; the role of probabilistic performance assessment. *Risk Analysis* (to appear).

Kaplan, S. 1991. The general theory of quantitative risk assessment. In Y.Y. Haimes, D. Moser, and E. Stakhiv, (Eds.), *Risk-Based Decision Making in Water Resources V*. New York: ASCE.

Kaplan, S. 1996. *An Introduction to TRIZ, the Russian Theory of Inventive Problem Solving.* Southfield, MI: Ideation International, Inc.

Kaplan, S. 1997. The words of risk analysis. *Risk Analysis* 17(4): 407-417.

North, D.W. 1960. A tutorial introduction to decision theory. *IEE Transactions on Systems Science and Cybernetics* SSC 4(3): 200-213.

A Fuzzy Rule-Based Model to Link Circulation Patterns, ENSO, and Extreme Precipitation

Agnes Galambosi,[1] Lucien Duckstein,[2] Ertunga Ozelkan,[3] and Istvan Bogardi[4]

Abstract

The purpose of this paper is to develop a fuzzy rule-based model (FRBM) to analyze local monthly extreme precipitation events conditioned on macrocirculation patterns (CPs) and El Niño/Southern Oscillation (ENSO). A case study in Arizona is presented to illustrate the methodology. The input variables of the FRBM consist of the monthly CPs and lagged Southern Oscillation Index (SOI) data; the output of the model is an estimate of local extreme precipitation. The daily CPs have been previously defined over the western United States by an automated clustering method, namely, principal component analysis coupled with K-means clustering. The SOI data has been divided into three fuzzy groups, representing the El Niño, normal, and La Niña years. The local extreme precipitation consists of monthly total precipitation from different Arizona precipitation stations. After defining and analyzing the basic properties of the extremes, a fuzzy rule-based model is constructed, and the results are interpreted and then compared with that of a

---

[1]Graduate student, Department of Systems and Industrial Engineering, University of Arizona, Tucson, AZ 85721
[2]Professor, Department of Systems and Industrial Engineering, University of Arizona, Tucson, AZ 85721
[3]Graduate student, Department of Systems and Industrial Engineering, University of Arizona, Tucson, AZ 85721
[4]Professor, Civil Engineering Department, University of Nebraska, Lincoln, NE 68588

multivariate linear regression model (MLRM). Using two goodness fit criteria, first the root mean squared error (RMSE) and then the correlation between the model results and the observed values, the FRBM is found to perform better than the MLR for capturing the properties of extreme precipitation events in Arizona.

Introduction

Although one can find several possible causes for local extreme precipitation events, this paper is devoted to the study of two causes: atmospheric circulation patterns (CPs), and El Niño/Southern Oscillation (ENSO). The purpose of this paper is to establish a relationship between ENSO events, large-scale CPs, and local extreme precipitation by using a fuzzy-rule-based model (FRBM), and then a multiple linear regression model (MLRM). Several studies (cited below) have shown that a linkage exists not only between CPs and local precipitation but also between ENSO and local precipitation. The use of information from both the CPs and ENSO (represented by the Southern Oscillation Index, SOI) is expected to provide better results than using just the CPs as inputs to downscale extreme precipitation. After establishing the relationship, it is possible to extend these models to the $2 \times CO_2$ scenarios; thus the possible impacts of ENSO on extreme precipitation in Arizona under climate change can be found. This can make it possible to improve the downscaling results for climate change studies. The present paper will give a basis for further climate change studies by specifying the connection between extreme events and their two possible causes as mentioned above.

The atmospheric CPs play an important role in climate research. For example, recently Woodhouse (1997) applied principal component analysis to winter atmospheric circulation patterns in the US Sonoran desert. Schubert and Henderson-Sellers (1997) downscaled local daily temperature extremes from CPs in Australia; Cavazos (1997) used an artificial neural network model for downscaling large-scale circulation to local winter rainfall in Mexico; Pesti et al. (1996) built a fuzzy-rule-based model for drought assessment; and Ozelkan et al. (1996) established a relationship between monthly CPs and precipitation by using fuzzy logic and regression approaches. Several stochastic models of daily precipitation have also been developed. They include, for example, Hay (1991), Bardossy and Plate (1992), Bogardi et al. (1993), and Bartholy et al. (1995).

The phenomenon called El Niño was initially recognized as a local climate anomaly of the eastern equatorial Pacific region, but today it is also associated with the Southern Oscillation, affecting climate events worldwide through teleconnection patterns. We now are aware of effects of ENSO on droughts in Indonesia, India, and Australia; floods in southern Brazil, Peru, and Ecuador; and epidemics in South America - just to name a few of them. To understand the physics of ENSO, several studies have been performed. Wang (1995) worked on understanding the transition from a cold to a warm state of ENSO, Njau (1996) developed a generalized theory of

ENSO and related phenomena, and Goddard and Graham (1997) described El Niño in the 1990s. Trenberth (1997) summarized the accomplishments and issues of short-term climate variations, and Hoerling et al. (1997) studied El Niño, La Niña, and the nonlinearity of their teleconnections. Chang (1997) looked at the ENSO extreme climate events and their impacts on Asian deltas, Lanzante (1996) studied the tropical SST lag relationships, and Chen et al. (1996) analyzed the interannual variation of atmospheric circulation associated with ENSO. Further details on the physical aspects and societal impacts of ENSO may be found in Glantz (1996).

The linkage between ENSO and precipitation events is also at the center of interest. Recently, Seleshi and Demaree (1995) studied Ethiopian rainfall variability and its linkage with SOI; Waylen et al. (1996b) reported a study relating Costa Rica's annual precipitation and the Southern Oscillation; Cullather et al. (1996) investigated the interannual variations of Antarctic precipitation related to ENSO; Kane (1997a) explored the relationships of ENSO with rainfall in Australia; Nicholson and Kim (1997) linked ENSO to African rainfall; Suppiah (1997) considered the connection between extremes of the Southern Oscillation (SO) and Sri Lanka rainfall; and Kane (1997b) compared ENSO, Pacific sea surface temperature (SST), and rainfall in various regions.

The relationship between ENSO and the climate of American regions also has been studied extensively. The following linkages have been investigated recently: inter-American rainfall and tropical Atlantic SST and Pacific variability (Enfield 1996); moisture conditions in the southeastern US and teleconnection patterns (Yin 1994); volcanic eruptions, ENSO, and US climate variability (Portman and Gutzler 1996); ENSO and Northern Plains precipitation and temperature anomalies (Bunkers et al. 1996); and ENSO and United States drought (Piechota and Dracup 1996).

Models of the relationship between ENSO and climate variables also cover a wide range. They include a vector time-domain approach to model Florida precipitation by Chu et al. (1995); a rotational empirical orthogonal function analysis for relationships between semi-arid Southern Africa rainfall and 700 hPa geopotential heights by Shinoda and Kawamura (1996); Monte Carlo simulations for ENSO-related precipitation and temperature across the Northern Plains by Bunkers et al. (1996); lag-relationships to predict global rainfall probability from SOI by Stone et al. (1996); lag cross-correlation studies for Costa Rica monthly precipitation interannual variability by Waylen et al. (1996a); and wavelet analysis of summer rainfall in China and India using SOI by Hu and Nitta (1996).

This paper is organized as follows: after describing the data, the extreme precipitation events in Arizona are defined and analyzed. The description and construction of the FRBM and MLRM follows. The results of the FRBM are then analyzed, interpreted, and compared to those of the MLRM.

Data

In order to establish a physical basis for the FRBM and the MLRM, this section is devoted to the description and analysis of the input and output variables of the FRBM and MLRM. The relationships between these input and output variables are then investigated.

The input variables of the FRBM and MLRM consist of clustered pressure height data and ENSO data. Forty-two years of daily observed 500 hPa level height data are used to describe the atmospheric CPs, defined at 35 grid points on a diamond grid influencing the southwestern US (Figure 1). The results of an automated, nonhierarchical method, namely principal component analysis (PCA) coupled with the K-means clustering technique, are used as an input to both the FRBM and the MLRM, having respectively, six, seven, seven, and eight types for winter (January-March), spring (April-June), summer (July-September), and fall (October-December). The description of the K-means clustering algorithm can be found in many statistical books and application papers as well, such as in MacQueen (1967) and Matyasovszky et al. (1993). Further details on the results in this area, such as the descriptions and hydrometeorological interpretations of the types are found in Galambosi et al. (1996).

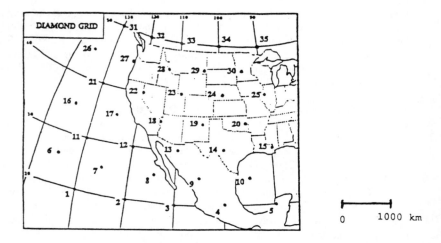

Figure 1. Diamond Grid over the Western United States for the Definition of CPs

In order to describe quantitatively the phenomenon called El Niño/Southern Oscillation, it is very common to use some indices which are functions of climatological variables. Here, the Southern Oscillation Index (SOI) is selected, which is the pressure difference between Tahiti and Darwin:

$$SOI = p(\text{Tahiti}) - p(\text{Darwin}) \qquad (1)$$

where p denotes the surface pressure (Clarke and Li 1995). Although nearly 130 years of historical SOI are available, only that period is utilized here which also has both CP and Arizona precipitation data available (1949-1989). Since the SOI is available on a monthly rather than on a daily basis, we either have to restrict this model to a monthly resolution, or find a way to distribute the monthly SOI data into smaller temporal scales. Here, we analyze the model on a monthly basis.

The output variable of the models is local precipitation in Arizona stations. Daily precipitation data can be selected from about 300 Arizona stations with available observations daily. The data stations have been selected on the basis of data availability, data homogeneity, representation of Arizona subclimates, and to provide a good spatial distribution. Figure 2 shows the eight Arizona stations selected ·or this study.

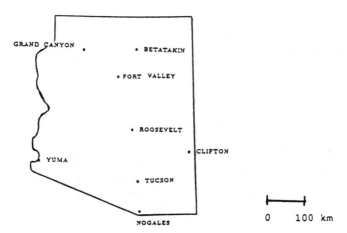

Figure 2. Arizona Precipitation Stations Selected for this Study

Although there are several ways to define extreme events, for simplicity here we chose the "threshold cut method": extreme data are defined on the basis of a threshold value, above which we call the data points extreme. Since the resulting set of extremes depends on the threshold value, a sensitivity analysis will be performed by changing their values. Note that here only extreme precipitation events are discussed, since drought (lack of precipitation) should be treated in a different way.

Results of a Preliminary Data Analysis

An example of two stations (Grand Canyon and Nogales) and two months (August and December) have been selected to illustrate the methodology. Four threshold values for the definition of extreme events have been chosen as 0.0, 0.2, 0.4, and 0.6 on a normalized scale, for sensitivity analysis purposes. Note that choosing 0.0 as a threshold means that the model is applied to the complete data base. Since we also include the "normal" precipitation data in the analysis, this makes it possible to generalize the approach and compare it with the model developed for extreme precipitation.

(a)   (b)

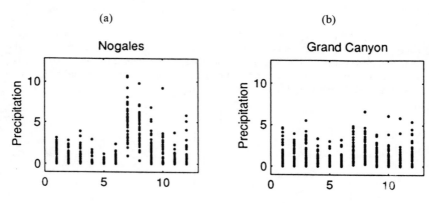

Figure 3. The Distribution of Precipitation in
(a) Nogales and (b) Grand Canyon

Figure 3 shows the distribution of precipitation at the selected stations of Nogales and Grand Canyon. The highest precipitation values occur in the second half of the year in Nogales, with the maximum in July and August (summer monsoon); high values also occur in September and December. At Grand Canyon station, the distribution exhibits smaller variability; the difference between the wet and dry months is not as strong as in Nogales. High precipitation occurs not only in July and August, but also in December, January, and March.

Figure 4 shows the lag correlations from 0 to 12, threshold 0.0, August and December, at both stations. We choose those lags from the figures which show high absolute value correlation between SOI and precipitation. The term "lag k", $k = 0$, . . ., 12 means that precipitation in a given month was plotted against SOI which occurred k months before. In August in Nogales, lags 1, 2, 3, and 7 have high correlation values, whereas in December, the lags are 2, 4, 5, and 6. The lag correlations will be used as inputs for the RBM.

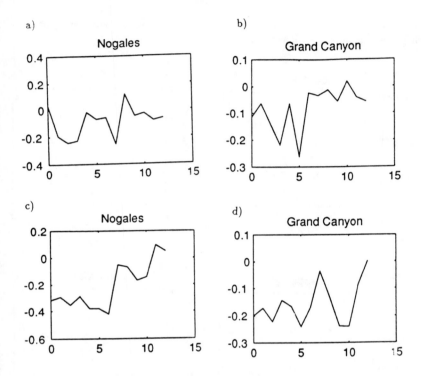

Figure 4. Lag Correlations from 0 to 12, Threshold 0.0:
a) August, Nogales; b) August, Grand Canyon;
c) December, Nogales; d) December, Grand Canyon

Figure 5 shows an example of the precipitation vs. SOI index for lag zero in August and for lag 2 in December, for both stations. The lag 2 case in December clearly shows that there is a good relationship between the local precipitation events and El Niño events at both stations, since high values of precipitation are associated with negative values of SOI. This was also indicated by the corresponding high values of correlation in Figure 4. In August for lag zero, the relationship is not that well defined, as we can see from both Figures 4 and 5. This clearly shows that factors other than SOI must play some causal role for precipitation in August.

Figure 6 shows several histograms of the CPs. Examining these figures provides the clue in a given month, at a given station, for a given threshold, which one of the CPs may cause precipitation. Part a) of Figure 6 shows the histograms for threshold 0.0 ("normal", or unconditional, events) in August and December, part b) of the figure corresponds to threshold 0.6 for both stations in August, and part c) to threshold 0.6 for both stations in December.

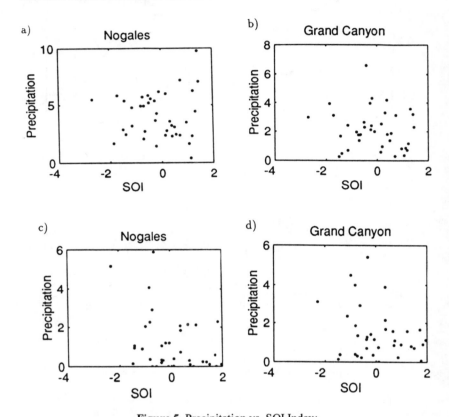

Figure 5. Precipitation vs. SOI Index:
a) Lag Zero, August, Nogales; b) Lag Zero, August, Grand Canyon;
c) Lag 2, December, Nogales; d) Lag 2, December, Grand Canyon

The histograms for unlike thresholds are different, meaning that it is not the same CPs that cause the "normal" and the "extreme" precipitation events. For example, for threshold 0.0 in August, CP 2 has the highest frequency, whereas for threshold 0.6, CP 7 occurred most, which means that this is an extreme precipitation-causing type. Also, the frequency distributions for the two selected stations and the two selected months appear to be different, indicating temporal and spatial climatic differences, as one would expect.

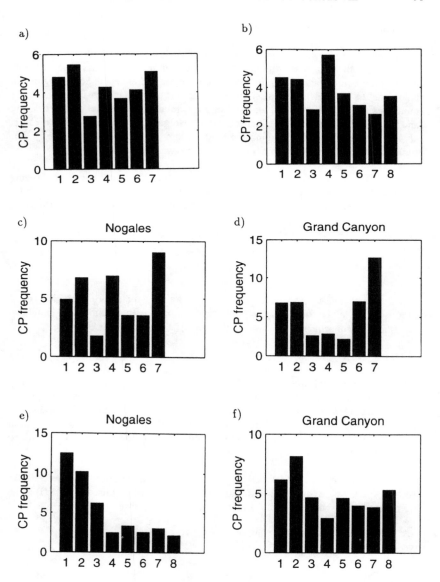

Figure 6. Histograms for a) Threshold 0.0, August; b) Threshold 0.0, December; c) Threshold 0.6, August, Nogales; d) Threshold 0.6, August, Grand Canyon; e) Threshold 0.6, December, Nogales; f) Threshold 0.6, December, Grand Canyon

Methodology: Fuzzy Rule-Based Model

Fuzzy logic, which was introduced by Zadeh (1965), provides a framework for dealing with vague and/or linguistic information in modeling. Fuzzy logic accepts overlapping boundaries of sets. In other words, instead of elements belonging or not belonging to a given set, partial membership in any set is possible. There are numerous techniques proposed for building fuzzy models (Nomura et al. 1992; Wang and Mendel 1992a, 1992b; Ozelkan et al. 1996; Higgins and Goodman 1994; Pesti et al. 1996). Here, we are describing the methodology of an FRBM very briefly. For further details on this subject, refer to, e.g., Bardossy and Duckstein (1995).

A fuzzy number is described by a membership function, $\mu_{A_{i,n}}(a_n)$, showing "how much" each observed data point a_n belongs to the set $A_{i,n}$. In this work, triangular fuzzy numbers (piecewise linear functions) are used for the calculations:

$$\mu_A(x) = \begin{cases} \frac{x - x^-}{x' - x^-} & x^- \le x \le x \\ \frac{x^+ - x}{x^+ - x'} & x' \le x \le x^+ \\ 0 & \text{elsewhere} \end{cases}$$

In general, a fuzzy rule consists of a set of explanatory variables called premises $A_{i,n}$ given in the form of fuzzy numbers with membership functions $\mu_{A_{i,n}}(a_n)$. The consequence B_i also is in the form of a fuzzy number:

$$\text{If } a_1 \in A_{i,1} \odot a_2 \in A_{i,2} \odot \dots \odot a_N A_{i,N} \text{ then } b \in B_i \tag{2}$$

where $\odot$ is a logical operator specified according to the application. Usually, rules are formulated using "AND" or "OR" operators. In contrast to ordinary (crisp) rules, fuzzy rules allow partial and simultaneous fulfillment of rules. This means that instead of the usual case when a rule is applied or is not applied, a partial applicability is also possible. We have summarized below the fuzzy rule-based framework used in this study.

Let $\{(a_1(t), \dots, a_N(t), b(t)); t = 1, \dots, T\}$ denote the set of observed data (training set), where $(a_1(t), \dots, a_N(t))$ is the input vector of premise values, $b(t)$ is the corresponding response at time t, and the total number of observations is T.

Step 0: Split sampling

Divide the input-output database into two parts. The first part will be used for calibration, to establish the relationship, and the second part for validation purposes, to test the performance of the model for prediction.

Step 1: Divide input-output space into fuzzy regions

Let us assume that the domain of each premise value a_n, $n = 1, \ldots, N$ and consequence b can be partitioned into a number of overlapping fuzzy sets: $\hat{P}_n = \{\hat{P}_{n,l}; l = 1, \ldots, L_n\}$, where L_n is the number of partitions and $\hat{P}_{n,l}$ is the fuzzy partition l for premise n. The fuzzy number $A_{i,n}$ is thus an element of these partitions: $A_{i,n} \in \hat{P}_n$. Similarly, let $\hat{P}_b$ be the set of partitions for consequence b, with the number of partitions L_b, so $B_i \in \hat{P}_b$.

Step 2: Generate fuzzy rules from given data

First, for a given data point $(a_1(t), \ldots, a_N(t), b(t))$; $t = 1, \ldots, T$, the memberships corresponding to the fuzzy regions are determined. Second, we assign the data point to the fuzzy sets with the maximum membership values. The input-output data thus yields a rule. For example, if $(a_1(1), a_2(1), b(1))$ would be assigned to "Low, High, Medium", respectively, the following fuzzy rule would be obtained:

$$\text{If} \quad a_1 \text{ is Low} \odot a_2 \text{ is High} \quad \text{then} \quad b \text{ is Medium} \tag{3}$$

If this rule is not previously encountered, we add it to the rule bank, which will be used later for prediction purposes. This procedure is repeated for each input-output data pair.

Step 3: Rule fulfillment

Once the fuzzy partitions have been determined, the degree of fulfillment of any rule can be calculated using the $\odot$ logical operator (Equation (2)) taken as the "AND" operator (Bardossy and Disse 1993).

$$\nu_i(t) = \prod_{n=1}^{N} \mu_{A_{i,n}}(a_n(t)) \tag{4}$$

The choice of the logical operator depends on the physics of the problem. In this paper, we have assumed that the "AND" operator explains satisfactorily the relationship between SOI, monthly CPs, and local extreme precipitation.

Step 4: Assigning a weight to each rule

To show which portion of the calibration data is explained by a given rule, one can weigh each rule by

$$\xi_i = \sum_{t=1}^{T} \nu_i(t) \, \mu_{Bi}(b(t)) \tag{5}$$

where ξ_i is the weight associated with rule i. At this step, the rules can be eliminated if desired by setting a threshold considered to be inapplicable. Note that in the present study $\xi^* = 0$ has been selected, meaning that all rules which are constructed from the observed data are retained.

Step 5: Calibration using the rule-based system

After the construction of the rules, the response of the fuzzy rules to a given input vector $(a_1(t), \ldots, a_N(t))$ can be computed using the training set. For this purpose, the fulfillment of each rule is calculated using Equation (4).

Step 6: Defuzzification

The response is then expressed as a real number as follows:

$$\hat{b}(t) = \frac{\sum_i \nu_i(t)\, \xi_i\, \beta_i^\circ}{\sum_i \nu_i(t)\, \xi_i} \tag{6}$$

where $\hat{b}(t)$ is the estimated value of $b(t)$ and β_i° is the most likely value of the consequence B_i as described in Step 1.

Step 7: Validation of the model

The output for each data point in the second part of the database is estimated by using the rules of the rule-based system established in the calibration.

The model developed here can be made adaptive if after each prediction, Steps 2 to 4 of the above-described algorithm are repeated as new data are observed. This way, if the rule explained by a new observation is different from the rules in the rule bank, a new rule is learned from the model and the rule weights are updated.

Methodology: Multivariate Linear Regression

In the case of a linear regression, we suppose that the true relationship between the dependent and independent variables is linear, so we assume that each observation y can be described as a linear function of the regressor variables $x_1, x_2, \ldots, x_n$:

$$Y = \beta X + \epsilon$$

where ϵ is a random error. When testing the model goodness statistically, it must satisfy a set of hypotheses often violated in practice, in particular that of no correlation between the input variables and homoscedasticity. The model parameters can be estimated by using the method of least squares. Further details of regression can be found, for example, in Chatterjee and Price (1991).

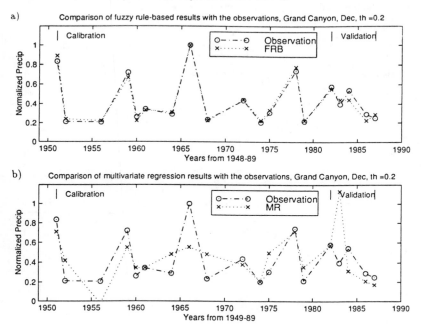

Figure 7. Comparison of Model Results with Observations: Grand Canyon, December, Threshold = 0.2: a) FRBM, b) MLRM

Application and Results

As mentioned before, a split sampling approach has been used to calibrate and validate the model. The number of years chosen for validation varies between 0 to 8 years because it depends on the total record length. For example, eight years of validation is used for all 0.0 threshold cases, whereas only four years have been validated for threshold 0.2, December cases. Then the frequency distribution of the CPs were obtained (where the CPs are the result of a principal component analysis coupled with K-means clustering by Galambosi (1996)). These and the SOI values of highest absolute value lag-correlations have been selected as inputs to both the FRBM and the MLRM. Thus the number of inputs changes for each month; for example, for Nogales precipitation in August, we have selected four values for the SOI lags (1, 2, 3, and 7), and the number of CPs in summer is 7. Furthermore, the number of CP and precipitation fuzzy partitions has been optimized so as to

minimize the root mean squared error (RMSE). Finally, FRBM and MLRM results
are compared, based on the RMSE and the correlation of the observed and model
values in both the calibration and validation periods.

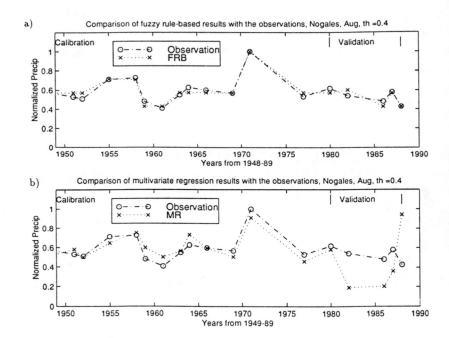

Figure 8. Comparison of Model Results with Observations, Nogales,
August, Threshold = 0.4: a) FRBM, b) MLRM

Figures 7 and 8 show comparisons of the model results with the observations
for both FRBM and MLRM. Two examples, Grand Canyon in December, threshold
0.2, and Nogales in August, threshold 0.4, are selected to illustrate the results. For
both cases illustrated here, the FRBM is superior to the MLRM, giving better fit not
only for the calibration but also for the validation period.

The overall comparison of both methods is shown more precisely in Figures 9
and 10. From all figures, it is clear that the FRBM (indicated as a dashed line) is
superior to the MLRM (shown as a dotted line) in each and every case when both the
CPs and the SOI data have been used as input variables, independently of the month,
location, or threshold. Using high threshold values, the MLRM usually gets worse
with both indicators. This can be explained by the fact that as a lower number of data
points are used for the regression, it becomes more unstable. Sometimes there are

fewer data points than the number of input variables, leaving no "degrees of freedom" for a valid regression analysis. On the other hand, even when data are not sufficient for the MLRM, the FRBM performs well, yielding robust results.

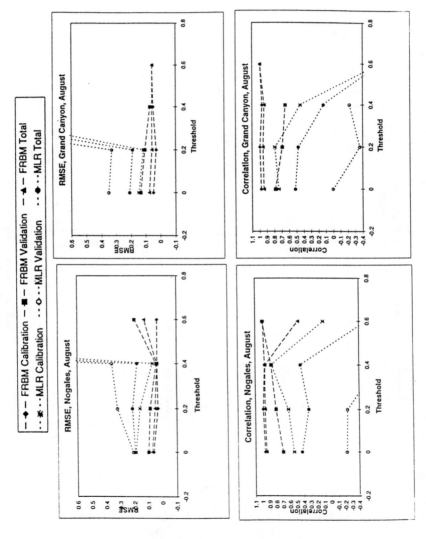

Figure 9. Values of RMSE and Correlation for August
in Nogales and Grand Canyon

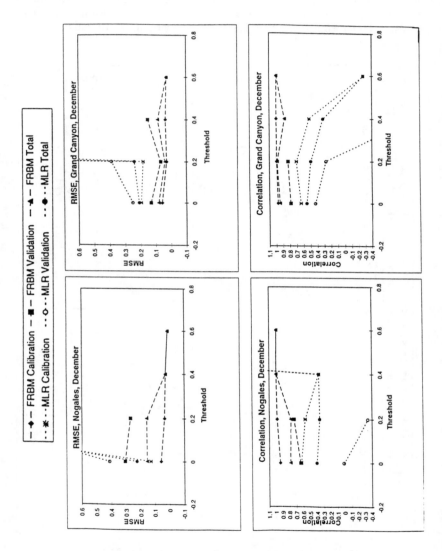

Figure 10. Values of RMSE and Correlation for December
in Nogales and Grand Canyon

Summary and Conclusions

This study examines the possible effects of El Niño and CPs on Arizona extreme precipitation. The methodology proposed here uses atmospheric CPs of the 500 hPa level and SOI of various time lags as inputs in order to predict local extreme precipitation events. Unlike some previous downscaling methods, here not only CPs are used for predicting a local hydroclimatological variable, but also the effects of ENSO are taken into consideration by using SOI as input to the model. Since El Niño plays an important role in local weather events via teleconnections, better results are obtained here than by using the CPs alone.

The findings of this study can be summarized as follows:

1. The FRBM appears to yield good predictions for the given area for both the"normal" and the extreme precipitation cases.

2. In terms of the RMSE and the correlation between observed and modeled values, FRBM is superior to MLR.

3. The FRBM is capable of giving good results even when there are insufficient data for MLR.

4. The FRBM gives more robust results in both the calibration and the validation periods than the MLRM.

Future work might include the expansion of the model to more ENSO variables, not only the SOI. Also, the use of daily rather than monthly quantities is planned by disaggregating the SOI data. Since the FRBM works very well in establishing the relationship between ENSO, CPs, and local precipitation, further studies of extreme events are possible for downscaling under climate change by using $2xCO_2$ GCM output scenarios.

Acknowledgments

This paper has been partially supported by the US National Science Foundation under grants CMS-9613654 and CMS-9614017. The authors also wish to thank Judit Bartholy and Istvan Matyasovszky for their help.

References

Bardossy, A., and E. Plate. 1992. Space-time model for daily rainfall using atmospheric circulation patterns. *Water Resources Research* 28: 1247-1259.

Bardossy, A., and M. Disse. 1993. Fuzzy rule-based models for infiltration. *Water Resources Research* 29(2): 373-382.

Bardossy, A., and L. Duckstein. 1995. *Fuzzy Rule-Based Modeling in Geophysical, Economic, Biological, and Engineering Systems.* Boca Raton, FL: CRS Press.

Bartholy, J., I. Bogardi, and I. Matyasovszky. 1995. Effect of climate change on regional precipitation in Lake Balaton watershed. *Theoretical and Applied Climatology* 51(4): 237-250.

Bogardi, I., I. Matyasovszky, A. Bardossy, and L. Duckstein. 1993. Application of a space-time stochastic model for daily precipitation using atmospheric circulation patterns. *Journal of Geophysical Research* 98(D9): 16653-16667.

Bunkers, M.J., J.R. Miller, and A.T. Degaetano. 1996. An examination of El Niño/La Niña-related precipitation and temperature anomalies across the northern plains. *Journal of Climate* 9(1): 147-160.

Cavazos, T. 1997. Downscaling large-scale circulation to local winter rainfall in northeastern Mexico. *International Journal of Climatology* 17(10): 1069-1082.

Chang, W.Y.B. 1997. ENSO-extreme climate events and their impacts on Asian deltas. *Journal of the American Water Resources Association* 33(3): 605-614.

Chatterjee, S., and B. Price. 1991. *Regression Analysis by Example.* New York, NY: John Wiley and Sons.

Chen, T.C., M.C. Yen, J. Pfaendtner, and Y.C. Sud. 1996. A complementary depiction of the interannual variation of atmospheric circulation associated with ENSO events. *Atmosphere-Ocean* 34(2): 417-433.

Chu, P.S., R.W. Katz, and P. Ding. 1995. Modeling and forecasting seasonal precipitation in Florida – a vector time-domain approach. *International Journal of Climatology* 15(1): 53-64.

Clarke, A.J., and B. Li. 1995. On the timing of warm and cold El Niño/Southern Oscillation events. *Journal of Climate* 8: 2571-2575.

Cullather, R.I., D.H. Bromwich, and M.L. Vanwoert. 1996. Interannual variations in Antarctic precipitation related to El Niño/Southern Oscillation. *Journal of Geophysical Research – Atmospheres* 101(D14): 19109-19118.

Enfield, D.B. 1996. Relationships of inter-American rainfall to tropical Atlantic and Pacific SST variability. *Geophysical Research Letters* 23(23): 3305-3308.

Galambosi, A., L. Duckstein, and I. Bogardi. 1996. Evaluation and analysis of daily atmospheric circulation patterns of the 500 hPa pressure field over the southwestern USA. *Atmospheric Research* 40: 49-76.

Glantz, M.H. 1996. *Currents of Change, El Niño's Impact on Climate and Society.* Cambridge, UK; New York, NY: Cambridge University Press.

Goddard, L., and N.E. Graham. 1997. El Niño in the 1990s. *Journal of Geophysical Research – Oceans* 102(C5): 10423-10436.

Hay, L.E. 1991. Simulation of precipitation by weather type analysis. *Water Resources Research* 27: 493-501.

Higgins, C. M., and R.M. Goodman. 1994. Fuzzy rule-based networks for control. *IEEE Transactions on Fuzzy Systems* 2(1): 82-88.

Hoerling, M.P., A. Kumar, and M. Zhong. 1997. El Niño, La Niña, and the nonlinearity of their teleconnections. *Journal of Climate* 10(8): 1769-1786.

Hu, Z.Z., and T. Nitta. 1996. Wavelet analysis of summer rainfall over North China and India and SOI using 1891-1992 data. *Journal of the Meteorological Society of Japan* 74(6): 833-844.

Kane, R.P. 1997a. On the relationship of ENSO with rainfall over different parts of Australia. *Australian Meteorological Magazine* 46(1): 39-49.

Kane, R.P. 1997b. Relationship of El Niño/Southern Oscillation and Pacific sea surface temperature with rainfall in various regions of the globe. *Monthly Weather Review* 125(8): 1792-1800.

Lanzante, J.R. 1996. Lag relationships involving tropical sea surface temperatures. *Journal of Climate* 9(10): 2568-2578.

MacQueen, J. 1967. Some methods for classification and analysis of multivariate observations. In *5th Berkeley Symposium on Mathematical Statistics and Probability* 1: 281-297.

Matyasovszky, I., I. Bogardi, A. Bardossy, and L. Duckstein. 1993. Estimation of local precipitation statistics reflecting climate change. *Water Resources Research* 29: 3955-3968.

Nicholson, S.E., and E. Kim. 1997. The relationship of the El Niño/Southern Oscillation to African rainfall. *International Journal of Climatology* 17(2): 117-135.

Njau, E.C. 1996. Generalized theory of ENSO and related atmospheric phenomena. *Renewable Energy* 7(4): 339-352.

Nomura, H., I. Hayashi, and N. Wakami. 1992. A learning method of fuzzy inference rules by descent method. *IEEE Proceedings: International Conference on Fuzzy Systems*, San Diego, CA.

Ozelkan, E.C., F. Ni, and L. Duckstein. 1996. Relationship between monthly atmospheric circulation patterns and precipitation: fuzzy logic and regression approaches. *Water Resources Research* 32(7): 2097-2103.

Pesti, G., B.P. Shrestha, and L. Duckstein. 1996. A fuzzy rule-based approach to drought assessment. *Water Resources Research* 32(6): 1741-1747.

Piechota, T.C., and J.A. Drakup. 1996. Drought and regional hydrologic variation in the United States – associations with the El Niño/Southern Oscillation. *Water Resources Research* 32(5): 1359-1373.

Portman, D.A., and D.S. Gutzler. 1996. Explosive volcanic eruptions, the El Niño/Southern Oscillation, and US climate variability. *Journal of Climate* 9(1): 17-33.

Schubert, S., and A. Henderson-Sellers. 1997. A statistical model to downscale local daily temperature extremes from synoptic-scale atmospheric circulation patterns in the Australian region. *Climate Dynamics* 13(3): 223-234.

Seleshi, Y., and G.R. Demaree. 1995. Rainfall variability in the Ethiopian and Eritrean highlands and its links with the Southern Oscillation Index. *Journal of Biogeography* 22(4-5): 945-952.

Shinoda, M., and R. Kawamura. 1996. Relationships between rainfall over semi-arid Southern Africa, geopotential heights, and sea surface temperatures. *Journal of the Meteorological Society of Japan* 74(1): 21-36.

Stone, R.C., G.L. Hammer, and T. Marcussen. 1996. Prediction of global rainfall probabilities using phases of the Southern Oscillation Index. *Nature* 384(6606): 252-255.

Suppiah, R. 1997. Extremes of the Southern Oscillation phenomenon and the rainfall of Sri Lanka. *International Journal of Climatology* 17(1): 87-101.

Trenberth, K.E. 1997. Short-term climate variations – recent accomplishments and issues for future progress. *Bulletin of the American Meteorological Society* 78(6): 1081-1096.

Wang, B. 1995. Transition from a cold to a warm state of the El Niño/Southern Oscillation cycle. *Meteorology and Atmospheric Physics* 56(1-2): 17-32.

Wang, L., and J.M. Mendel. 1992a. Generating fuzzy rules by learning from examples. *IEEE Transactions on Systems, Man, and Cybernetics* 22(6): 1414-1427.

Wang, L., and J.M. Mendel. 1992b. Fuzzy-basis functions, universal approximation, and orthogonal least-squares learning. *IEEE Transactions on Neural Networks* 3(5): 807-814.

Waylen, P.R., C.N. Caviedes, and M.E. Quesada. 1996a. Interannual variability of monthly precipitation in Costa Rica. *Journal of Climate* 9(10): 2606-2613.

Waylen, P.R., M.E. Quesada, and C.N. Caviedes. 1996b. Temporal and spatial variability of annual precipitation in Costa Rica and the Southern Oscillation. *International Journal of Climatology* 16(2): 173-193.

Woodhouse, C.A. 1997. Winter climate and atmospheric circulation patterns in the Sonoran desert region, USA. *International Journal of Climatology* 17(8): 859-873.

Yin, Z.Y. 1994. Moisture conditions in the southeastern USA and teleconnection patterns. *International Journal of Climatology* 14(9): 947-967.

Zadeh, L.A. 1965. Fuzzy sets. *Information and Control* 8(3): 338-353.

Managing Uncertainty

William D. Rowe[1]

Abstract

There is uncertainty in all aspects of life. In business, this uncertainty particularly manifests itself in system performance, risks, cost estimation, and benefits. How can these uncertainties be managed to the advantage of the enterprise? This paper provides a strategic approach to managing uncertainties in a decision-making environment. It is based upon four processes: 1) a means for the analytical decomposition of sources of uncertainty, making them visible to both the analyst and the manager; 2) a method for rapidly capturing existing information for all parameters from a heterogeneous array of sources and using these in decision models; 3) an information value analysis process (IVAP) for developing a cost-effective strategy for acquiring new information; and 4) a means to present results and their implications on-line in an easily understandable "what-if" response mode.

Introduction

We live in an uncertain world, and seek to reduce this uncertainty to a manageable level by acquiring information about the physical world and the people with whom we interact.

As individuals, we have learned to enhance our rather limited sensory perceptions to some degree, particularly sight and sound. Scholarly information has been pooled over time and populations to provide libraries and other repositories of knowledge. Most of this information is qualitative; it resides in books, databases, and other media, and is imparted through education as well as data interchange systems.

---

[1]President, Unlimited Assurance Holdings, Inc., 726 Battery Place, Alexandria, VA 22314

Quantitative data is obtained through measurements and is then aggregated into information for use in the scientific, business, mathematical, sociological, and technology disciplines. It is impossible to gain all information on any subject, and acquiring information can be costly.

This is no surprise to the enterprise where decisions involving system performance, risks, cost estimation, and benefits are made all the time. This paper provides a strategic approach to managing uncertainties in a decision-making environment. The approach is based upon four processes, providing:

1) a means for the analytical decomposition of sources of uncertainty to make them more understandable and visible to both the analyst and the manager;
2) a method for rapidly capturing existing information for all parameters from a heterogeneous array of sources and using these in decision models;
3) an information value analysis process (IVAP) for developing a cost-effective strategy for acquiring new information; and
4) a means to present results and their implications on-line in an easily understandable "what-if" response mode.

This strategy is outlined below.

Types of Uncertainty

Uncertainty is present in all decisions we make. Usually there is more that we can learn, no matter how much we already know. This uncertainty comes in four forms or types:

1) Temporal – uncertainty of future or past states,
2) Structural – uncertainty due to complexity,
3) Metrical – uncertainty in measurement, and
4) Translational – uncertainty in explaining results (Rowe 1994).

All four types may occur in any situation, but one or more usually dominates. Although these types are not necessarily independent, the nature of each is quite different. Each type can be addressed separately, and then their interaction may be examined, as shown in Table 1.

Table 1. Parameters of the Types of Uncertainty

Table 1. Parameters of the Types of Uncertainty

UNCERTAINTY CLASS	UNCERTAINTY SOURCE	DISCRIMINATION PARAMETER	VALUATION PARAMETER	METHODS USED
TEMPORAL	Future	Probability	Luck	Prediction
TEMPORAL	Past	Historical Data	Correctness	Retrodiction
STRUCTURAL	Complexity	Usefulness/Likelihood	Confidence	Models
METRICAL	Measurement	Precision	Accuracy	Statistics
TRANSLATIONAL	Perspective	Goals/Values	Understanding	Communication

Since uncertainty is present in all decisions, and often cannot be reduced within the limits of time resources or knowledge, we must learn to manage it. The costs of ignoring uncertainty can be very high in terms of unwelcome surprises and poorly calculated risk-taking behavior. The challenge is to develop a useful approach to managing uncertainty in decisions that affect our everyday lives or policies at all levels.

Approaches to Managing Uncertainty

Uncertainty often cannot be reduced by acquiring new data, nor can one afford to obtain that data even if available. Alternatives to seeking better information are to manage the uncertainty with the information on hand or to use a very selective data acquisition program.

Uncertainty is addressed by at least five different broad approaches:

1. Ignore it.
2. Use margins of safety to provide contingencies.
3. Use contractual means to limit uncertainty and risk.
4. Buy insurance to spread the risks.
5. Directly understand and manage the uncertainties.

1. Ignoring Uncertainty

If we ignore uncertainty, the result is often a poor decision, leading to missed goals and opportunities and high costs. Moreover, the increased likelihood of unwelcome surprises can often lead to civil or even criminal liability due to negligence.

A good example is the problem of distinguishing "competent error" from either "negligent error" or "criminal behavior", particularly after an event has occurred. The following is an example: after examining the circumstances regarding past accidents, a 99-percent level of confidence is established that the accident will not occur under normal conditions. Should one be unlucky, the 1-percent event occurs, as the accident can occur on a random basis. Given that the accident distribution was correctly estimated, and that costs and the capability to lower the occurrence of the random event are unreasonable, this would be considered competent error. If the rate of occurrence is too high or unknown and steps have not been taken to reduce the accident rate, or if an incorrect sampling distribution was used, this situation would be considered negligent error. That is, steps have not been taken to reasonably limit the risks during design of the system. If improper information was knowingly used for establishing the design limits, this may even be considered criminal behavior.

Then there is the problem of finding accident causes before and after the fact. Before the fact, a decision maker decides that steps should be taken to limit risk to one chance in 1,000 at some level of confidence. This means that an accident may occur by chance alone one time out of 1000. After the fact, one not only looks at causes, but tries to place blame for the event with 20-20 hindsight. Who should be blamed? Should it be those who accepted the 1-in-1,000 level of risk, or should those working when the event occurred be blamed? For example, the Air Force has exonerated the officers managing the Saudi Arabia complex that was bombed recently, but Congress is still trying to place the blame. In a competently designed and operated system there will always be some small residual error. If one can't make the case that the accident is the result of competent error, it is usually considered to be negligent error. Establishing a means to differentiate among these types of errors is one way of avoiding the costs of litigation and negligence.

2. Using Margins of Safety to Provide Contingency

A common way to address uncertainty is to use margins of safety to provide for contingencies and provide "conservative" designs, where conservative implies erring on the "safe side". These margins of safety rarely come free. They imply a degree of over-design, resulting in increased costs. When used in more than one parameter, the margins of safety pyramid, resulting in extreme conservatism. When margins of safety are applied at low levels by each designer in the organization, again the pyramiding leads to extreme conservatism in decision options. This can force the decision makers at the top into constrained choices that are neither optimal nor cost-effective. This masking of uncertainties suppresses other, perhaps more desirable, alternatives, leading to missed opportunities.

Model choice uncertainty in assessing environmental threats provides a good example of this problem in regulatory practice, as illustrated in Table 2 (Rowe 1987). Six different types of models are required to estimate environmental risk, for three specific cases: a toxic organic compound, radio-iodine, and nitrogen dioxide. The range of uncertainty due to the choice of plausible, available alternative models is shown. Although these are qualitative estimates, in some cases they reflect the relative uncertainty in models due to the choice of which model to use. The bottom of the range represents situations with low variability, e.g., a diffusion model for a flat, even terrain. The higher values are for cases with high variability, e.g., a diffusion model operating on a hilly terrain with shear wind strata. When these ranges are aggregated, they do so multiplicatively, and since they are uniform distributions, their endpoints must be used. As can be seen from the multiplicative ranges in the table, the ranges vary from about one order of magnitude (factors of 10) to 14 orders of magnitude. The range for a toxic organic compound is from 5 orders of magnitude to 14.

With this kind of uncertainty, risk analysis cannot be very robust unless the means of reducing risk is so effective as to overcome these margins of safety. A good example is a hazardous waste facility with a destruction removal efficiency of 99.9999 percent (for PCBs and TCDDs) and a diffusion factor that results in dilution to 1 part in 10 million. This results in a total risk reduction of 13 orders of magnitude; enough to

Table 2. Model Choice Uncertainty in Environmental Assessment

MODEL	UNCERTAINTY FACTOR RANGE		
	Toxic Organic	*Radio-Iodine*	*Nitrogen Dioxide*
1. Source Term			
Averaging in Space	1.1 - 3	1.1 - 2	1.1 - 10
Averaging in Time	1.1 - 3	1.1 - 3	1.1 - 5
2. Pathways			
Diffusion Models	2 - 10	1.1 - 2	1.1 – 3
3. Metabolic Pathways and Fate			
Organ Intake Models	2 - 10	1.1 - 3	1.1 - 3
Distribution Models	2 - 4	1.1 - 2	1.1 - 2
Retention Models	2 - 4	1.1 - 1.5	1.1 - 1.5
4. Dose Estimate			
Exposure Time Profile	2 - 10	1.1 - 10	2 - 10
Maximum vs Individual Exposure	2 - 10	1.1 - 5	2 – 10
5. Dose Effect Relationship			
Extrapolation from Animal to Man			
Choice of Scaling Model	10 – 40	2 - 10	10 - 20
Metabolic Differences	2 - 100	1.1 - 1.5	2 - 5
Extrapolation From High to Low Dose			
Choice of Model (Cancer Case)*	10 - 1000	1 - 3	10 - 100
Margin of Safety (Threshold Case)*	10 - 1000	1 - 2`	2 – 10
6. Risk Estimation			
Individual Risk Estimate Case**			
Real vs. Hypothetical Individual	4 - 20	4 - 20	4 - 20
Population Risk Estimate**			
Integration vs. Averaging Models	2 - 10	2 -5	2 - 10
Multiplicative Ranges			
Low (Indiv.. Risk for Cancer)	3×10^5	2×10^1	3×10^2
High (Indiv. Risk for Cancer)	$.1 \times 10^{14}$	5×10^6	3×10^{10}

*, ** Use either, but not both.

overcome most of the uncertainty due to model choice. Conversely, a hazardous waste landfill cannot account for more than 3 orders of magnitude of risk reduction, and cannot overcome the effect of the aggregated margins of safety (Rowe 1987). Therefore, the use of extreme margins of safety can easily mask intelligent choices when these margins of safety pyramid.

3. Use Contractual Means to Limit Uncertainty and Risk

Legal contracts are used to limit one's responsibility for both competent and negligent error. This shifts the risk to others, usually at higher costs or inequitable distribution of risks. Greater uncertainty leads to higher negotiated costs.

Unfortunately, in many cases the negotiated legal requirements may have no relation to reality. An example of this occurred in the early seventies in the design of nuclear power plants. Facility designers were required to design within the parameters of a contractually designated maximum credible accident (Rasmussen 1975). Several years later, the Three Mile Island accident exceeded this hypothetical event. Suits to the plant operators are still not all settled. Chernobyl is another example where events exceeded all expectations.

Contracts have loopholes. These can be established explicitly by document wording or may be implicit when not specifically addressed. Contracts can be voided in the courts by judicial means, as well as through bankruptcy and criminal negligence and behavior.

4. Buy Insurance to Spread the Risks

Insurance and underwriting provide means to spread financial risks to an individual or organization over a much larger population. Premium charges are a function of the degree of uncertainty involved, the size and frequency of consequences, and last year's underwriting performance. However, limits are usually set for the extent and amount of coverage placed, with premiums reflecting these limits. Moreover, restrictions may be placed – for example, requiring management to use only scheduled airlines for travel, or mandating specified levels of training and licensing for employees in certain professions. Additionally, underwriters may balk at payout time, necessitating legal proceedings to obtain redress.

5. Directly Analyze and Manage the Uncertainties

Managing uncertainty requires acknowledging its existence in every decision, analyzing the impact of these uncertainties, having the means to help resolve the decisions involved, and taking remaining uncertainties into account. In brief, first it requires analyzing the uncertainties, and then applying the means to address them directly.

Reducing the uncertainties by acquiring missing information is one means of managing them if obtaining this information is cost-effective. However, irreducible uncertainties and limitations in time and resources are but a few reasons why this approach is not universally applicable. Risk analysis is basically another means of addressing uncertainties that involve future gambles, but it is only one of several approaches. Fuzzy-set theory is one approach to model uncertainty. This process assigns degrees of membership to alternate conditions. Multiobjective decision theory also addresses this type of uncertainty. Statistical methods are used to understand and manage measurement error.

Managing uncertainties directly requires a strategy for analyzing them and then using all the approaches mentioned above plus the many other available tools that can aid in reaching useful decisions. Methods 2 through 4 above also may be used strategically. The development and implementation of a strategic, all-encompassing approach is what is meant in the broadest sense by the term "managing uncertainty."

There are problems involved in directly managing uncertainties. First, the effort requires increased analysis with accompanying time and resources. Second, uncertainty is a difficult parameter to understand and to express, particularly to those who are not experienced or adept in handling uncertainties directly. Moreover, analyzing the uncertainties in a formal manner may make visible conditions that are politically, organizationally, and legally undesirable and even unacceptable. For example, political realities that focus on the short term may tend to suppress the impact of longer-term results, understanding the uncertainties may lead to politically contrary recommendations that otherwise remain invisible, and decisions made using recorded, quantified acceptable risk levels may provide the bases for future liability.

There is uncertainty in all parameters of a decision: performance, costs, benefits, risks, etc. Performance and benefits are often difficult to measure directly, particularly intrinsic benefits. Risks, the downside of a gamble involving probabilities, cannot be measured, only modeled (after the event occurs, the outcome is certain). The uncertainty in cost-estimation is often masked by accounting procedures which are designed for bookkeeping as opposed to estimation. The uncertainties in all these parameters must be addressed collectively.

A Generic Strategy for Managing Uncertainty

A generic strategy is proposed here for addressing decisions involving uncertainty. It consists of expanding the process into several steps, together with a variety of tools and methods for implementing each step.

Step 1. Develop a model for the decision and identify the parameters.

This involves developing a decision structure which includes: a) designing a model and perhaps, working up alternative model designs, b) identifying the uncontrollable parameters (states of nature), c) identifying the controllable parameters (alternative control conditions), and d) setting criteria for acceptance or rejection (value or utility functions – see Haimes 1980).

Step 2. Identify the existing heterogeneous data sources for the information on each parameter.

The following hierarchy is one way of categorizing data sources. It is based on the degree to which the data can be validated, ranging from empirically at the top to unvalidated data at the bottom.

1. Standard distribution

Empirical data validated by many different investigations and generally agreed upon as a standard. Example: the distribution of adult weight in the United States.

2. Empirical distribution

Empirical data validated for specific instances. Example: the distribution of arsenic concentrations in 20 samples for 10 wells.

3. Validated model

A model that has been validated empirically in repeated experiments. Example: determining altitude from local pressure instruments measuring surface pressure and humidity.

4. Unvalidated model

A model which has not been empirically validated. Example: inferring the effects on human cancer of low-dose medication from high-dose data on animals. Validation may or may not be possible.

5. Alternate models

Alternate unvalidated models that seem to be less reasonable, but possible, can be used in a decision process. Example: a threshold model for cancer. Alternate models may actually be superior to the primary model from differing perspectives.

6. Expert value judgment

In the absence of empirical data, the judgment of experts may be used. Since the information is unvalidated, the selection, framing, and bias of experts becomes dominant. Sometimes consensus may be important; otherwise, diversity may be important and

	retained. For example, should benign tumors that do not progress to malignancy be counted in mouse-exposure data?
7. Best-guess estimate	A form of presumably expert judgment based upon reasonable expectation and the framework in which the decisions are requested, but acknowledging the lack of integrity in the estimates at the outset.
8. Test case	Artificial values used to test models for their sensitivity, critical variables, and the value of additional information.

There are degrees between each step in the hierarchy, and one may add or delete levels, but the list provides a means to differentiate between sources of uncertainty (Richards and Rowe 1998).

Step 3. Rapidly capture the existing information from the heterogeneous data sources by taking "snapshots" of the data distributions for each parameter.

The use of range/confidence estimation techniques is one approach to acquiring this type of data rapidly. For details, see Richards and Rowe (1998).

Step 4. Using the captured data, run the model and evaluate the results. If a decision can be made with acceptable confidence using the initial data, accept the results, and stop, or else, go on to Step 5.

Step 5. Use an information value analysis process (IVAP) to develop a cost-effective plan for acquiring data to reduce the uncertainty to acceptable levels. This includes identifying the critical variables and focusing on them. Again, refer to Richards and Rowe (1998).

Step 6. Acquire the additional data according to the plan and rerun the model. Iterate until the decision can be made with acceptable confidence, or determine that the decision is "undecidable" in analytic form and stop.

Conclusions

By outlining the many approaches used to address uncertainty, the advantages and shortcomings of each can be compared. All of these approaches must be addressed simultaneously if an overall strategy is be developed. One such initial strategy for managing uncertainty is proposed. It consists of several steps that provide a rapid low-cost procedure for assessing the models used, the data, and the uncertainties analytically. Cost-effective data-acquisition strategies can be developed to achieve decision closure where there is adequate data or where further reduction of uncertainty is needed and is feasible. Although several approaches for carrying out this process have been mentioned as examples of the implementation of this strategy, it is premature to adopt any of them generically at this time. The message here is that an overall approach must be addressed and discussed, and a variety of methods and approaches developed, implemented, and then evaluated.

References

Haimes, Y.Y. 1980. Risk-benefit analysis in a multiobjective framework. In Y.Y. Haimes (Ed.), *Risk/Benefit Analysis in Water Resources Planning and Management*. New York, NY: Plenum Press.

Rasmussen, N. (Ed.). 1975. *The Reactor Safety Study*. Report WASH 1400, US Regulatory Commission, Washington, DC.

Richards, D. and W.D. Rowe. 1998. Decision making with heterogeneous sources of information. *Risk Analysis* (in publication).

Rowe, W.D. 1977. *An Anatomy of Risk*. New York: John Wiley Interscience.

Rowe, W.D. 1987. Alternative risk evaluation paradigms. In Y.Y. Haimes and E.Z. Stakhiv (Eds.), *Risk Analysis and Management of Natural and Man-made Hazards: Proceedings of the Third Conference*. New York: American Society of Civil Engineers.

Rowe, W.D. 1994. Understanding uncertainty. *Risk Analysis* 14(5): 743-749.

Risk of Regional Drought Influenced by ENSO

Rita Pongracz[1], Istvan Bogardi[2], Lucien Duckstein[3], and Judit Bartholy[4]

Abstract

El Niño/Southern Oscillation (ENSO), global atmospheric circulation patterns (CP), and regional drought are coupled using a fuzzy-rule-based methodology. A common drought index, the monthly Palmer Drought Severity Index (PMDI) is used to characterize drought conditions in the state of Nebraska. The methodology aims at reproducing the observed statistical properties of PMDI from large-scale synoptic information of daily CP and a common monthly ENSO index (SOI). Although PMDI is dependent statistically on the CP types and ENSO phases in Nebraska, the weak correlations preclude using regression analysis or stochastic modeling. To use the fuzzy-rule-based methodology, first fuzzy numbers are defined on the input and output variables: SOI, the monthly relative frequencies of the 10 CP types, and PMDI. Next, one part of the data is used to construct fuzzy rules. Then, PMDI is calculated with the fuzzy rules for the independent validation set. The fuzzy-rule-based model provided encouraging preliminary results when the weighted counting algorithm was used for rule construction.

Introduction

The purpose of this paper is to couple El Niño/Southern Oscillation (ENSO), global atmospheric circulation patterns (CP), and regional drought using a fuzzy-

[1]Graduate student, Department of Civil Engineering, University of Nebraska, W359 Nebraska Hall, Lincoln, NE 68588
[2]Professor, Department of Civil Engineering, University of Nebraska, W359 Nebraska Hall, Lincoln, NE 68588
[3]Professor, Department of Systems and Industrial Engineering, University of Arizona, Building 20, Tucson, AZ 85721
[4]Associate Professor, Department of Meteorology, Eotvos Lorand University, Ludovika ter 2, Budapest, H-1083, Hungary

114

rule-based methodology. A common drought index, the monthly Palmer Drought Severity Index (PMDI) is used to characterize drought conditions in the State of Nebraska. The methodology aims at reproducing the observed statistical properties of PMDI from large-scale synoptic information of daily CP and a common monthly ENSO index.

The term El Niño was originally used along the coasts of Ecuador and Peru to refer to unusually warm ocean currents appearing around Christmas-time and lasting for several months. Recently, El Niño has been defined as a warm oceanic water current flowing from west to east in the tropical Pacific Ocean. El Niño is inherently connected with Southern Oscillation (SO), a quasi-periodic variability of the atmospheric pressure of tropical atmosphere in the western and eastern parts of the Pacific Ocean. Trenberth and Shea (1987) defined SO as an air mass transport fluctuating between western and eastern hemispheres with centers over Indonesia and the tropical southeastern Pacific Ocean. Julian and Chervin (1978) and Rasmusson and Carpenter (1982) presented a more detailed description of SO. The two phenomena, El Niño and Southern Oscillation, are labeled with a common abbreviation: ENSO. The ENSO events, which also include La Niña, the opposite of El Niño, are the most important sources of year-to-year variability in climate over tropical Pacific areas, and also the second strongest climate signal on the planet, next to the seasons. A conference on the World Climate Research Program was held in Geneva from August 26-28, 1997 with participants from more than 100 countries. One of the most important topics was the actual strong El Niño event which was expected to persist throughout the year and into early 1998. According to this ENSO event, strongly anomalous precipitation conditions were observed over the islands of the central tropical Pacific, in central Chile, Argentina, and eastern Australia. However, ENSO has considerable effects on the weather in areas far from the tropical Pacific Ocean (Horell and Wallace 1981; Kiladis and Diaz 1989).

ENSO events frequently result in anomalous weather, especially in droughts or extremely wet conditions. Ropelewski and Halpert (1987) examined the relationship between ENSO and global scale distribution of precipitation. Four major areas affected by ENSO have been found in Australia, two in North America (southeastern and southwestern US), two in South America and India, and one in the Central American area. The western and southwestern US were studied especially intensively (Andrade and Sellers 1988; Redmond and Koch 1991; Woolhiser and Keefer 1993). Schonher and Nicholson (1989) analyzed California rainfall in the years from 1950 to 1982, a period which suffered 11 El Niño events. In southern California, 9 of the 11 El Niño events produced rainfall amounts above normal, and 8 of the 10 wettest years coincided with El Niño years. The magnitude of the influence of ENSO depends on the space-time distribution of Pacific sea surface temperature (SST), but the probability of severe rainfall is considerably larger in El Niño years. Presnell (1995) identified three preferred temperature and precipitation patterns in the contiguous United States prior to, during, and after ENSO events.

Horell and Wallace (1981) examined atmospheric phenomena associated with ENSO, with emphasis on vertical structure and teleconnections to mid-latitudes. The correlation between the tropical time series and northern hemisphere geopotential height fields exhibits well-defined teleconnection patterns: warm ENSO episodes tend to be accompanied by below-normal heights in the North Pacific and the southeastern United States and above-normal heights over western Canada.

To overcome the scale differences between large-scale synoptic forcing – here CP and ENSO, and local or regional scale climatic response – here a drought index, the large-scale data should be downscaled. The main challenge of downscaling is that contrary to obtaining integrated information from elements (upscaling), the reverse problem of obtaining local information from integrated information (downscaling) is more difficult. For this reason, various downscaling techniques utilize additional physically-based information in addition to empirical data. Giorgi and Mearns (1991) distinguish three main types of downscaling approaches. The empirical techniques (having the relatively simplest mathematical formulation) are based on an assumption that similar large-scale climate changes have similar local consequences independent of the factors causing the global change (Hansen et al. 1984). This hypothesis makes it possible to use proxy data from past eras to estimate the future regional climate change (Glantz 1988). Such estimates are based on the regional paleoclimate and the analogy between the paleo- and GCM-simulated global climates. A discussion of using paleos can be found in Covey (1995). Another type of downscaling includes nested modeling with regional atmospheric models. In this case, fine-resolution numerical models are nested in the grid of GCMs by using GCM outputs as boundary conditions of these meso-scale models (Dickinson et al. 1989). This technique requires substantial effort in terms of modeling and computer programming. A newly developed strategy is the use of time-slice GCM experiments when a high-resolution atmospheric GCM is forced by the boundary conditions for the atmosphere generated in low resolution coupled atmosphere-ocean GCM. However, no satisfactory long simulation is available to assess extremes. These difficulties, among others, motivate using the third type of downscaling approaches in the present paper, namely semi-empirical downscaling procedures which appear to be specific combinations of the previous two techniques.

The fuzzy-rule-based model is presented in four steps. First, the input and response data sets are described. Then, we show that although CP, ENSO, and regional drought are statistically dependent, regression analysis or stochastic modeling may be poor choices to identify the relationship between large-scale synoptic input and the drought response over the region considered. Next, the fuzzy-rule-based approach is developed and the application results are highlighted.

Input and Response Data

In the current investigation we used the following three data sets to develop the fuzzy-rule-based model.

1. The ENSO phenomena are represented by monthly values of the Southern Oscillation Index (SOI) (NOAA 1997), which is one of the most frequently used indicators of warm and cold ENSO events. SOI is defined as the difference between sea level air pressures in Tahiti (l7.5°S, 149.6°W) and Darwin (12.4°S, 130.9°E). The negative range of SOI implies a warm phase (or El Niño), while the positive range indicates a cold phase (or La Niña). Small absolute values (around 0) correspond to the case of a normal state.

2. Large-scale circulation patterns are represented by CP types for a 49-point grid which includes the US region and the adjacent Pacific areas and has Nebraska in its centroid. The CP-type system was derived by using k-mean clustering method for the principal components (Matyasovszky et al. 1993; Bartholy et al. 1994) and includes ten different circulation types. To form a daily CP catalogue for the years 1946-1994 we used daily geopotential heights of the 500 hPa level. The daily CP types can be used to define – among others – two types of possible inputs (or premises) for the fuzzy-rule-based model: the most frequent daily CP type for the month (non-fuzzy input), or the monthly frequencies of the 10 CP types (10 fuzzy inputs). This latter choice seems to be more appropriate since it provides more information about the large-scale circulation than the first one. Actually, in the present case, better results were obtained with the 10 fuzzy CP inputs. ·

3. Drought for Nebraska, the response variable, is represented by the modified Palmer drought severity index (PMDI or modified PDSI) (NOAA 1997) that has negative monthly values if there is drought: the greater the drought the bigger the absolute value of PMDI. Positive monthly values of PMDI imply wet conditions in the same manner. The PDSI is based on the principles of a balance between moisture supply and demand when man-made changes are not considered. This index indicates the severity of a wet or dry spell. The PDSI was modified by the National Weather Service Climate Analysis Center to obtain values which are more sensitive to the transition periods between dry and wet conditions (Heddinghause and Sabol 1991). In this paper, summer drought conditions are considered by using the half-year period from April through September.

Statistical Relationships between Datasets

The possible statistical relationships among the three datasets were analyzed pairwise by considering discrete categories of the Palmer Drought Index and SOI in the following way.

The Palmer Drought Index is considered very dry if the value is smaller than -3, dry between -3 and -1, normal between -1 and + 1, wet between + 1 and +3, and very wet if the value is bigger than +3. The ENSO categories are separated at the values of -1, between El Niño and normal, and +1, between normal and La Niña.

In the present study, no delay parameter was set between the values of drought index and other variables.

All the correlation coefficients between PMDI and SOI, PMDI and monthly relative frequencies of CP types, and SOI and monthly CP-relative frequencies are very low, smaller than 0.3 in absolute value.

In Figure 1 the empirical frequency distributions of drought conditions can be seen during the warm and cold ENSO phases. Chi-square tests for homogeneity showed that these empirical frequency distributions are different at the 0.02 significance level during the ENSO phases.

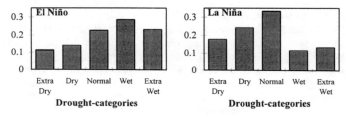

Figure 1. Empirical Relative Frequency Distributions of Drought Conditions during the Warm and Cold ENSO Phases

Next, the empirical frequency distributions of CP types prevailing during the five drought categories are determined. These statistical characteristics are also different at the 0.01 significance level. Figure 2 shows the relative frequencies of CPs for the two extreme drought conditions.

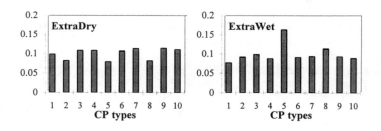

Figure 2. Empirical Relative Frequency Distributions of CP Types during Extreme Dry and Wet Conditions

Finally, the empirical frequency distributions of CP types during the three ENSO conditions are determined with different delay parameters from 0 to 6 months before. In Figure 3, the relative frequencies of CPs can be seen for the two opposite ENSO phases with 3 months delay.

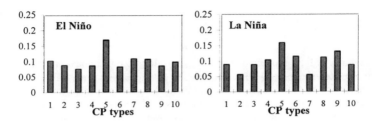

Figure 3. Empirical Relative Frequency Distributions of CP Types
during the Warm and Cold ENSO Phases

The above simple analysis indicates that the monthly Palmer Index in Nebraska is dependent on the CP types and ENSO phases. However, the correlation is so weak that a regression analysis would not lead to any meaningful results (Carlson et al. 1996). Then, stochastic modeling would also have difficulties, as shown by Pesti et al. (1996), because the prediction of the monthly Palmer drought index would require a stochastic vector input of 10 dependent random variables – the daily CP types in the 10 days preceding the day of prediction. In addition, these random variables are non-normal, and may not even have the same distribution. Another requirement would be the independence of input data, which is rarely satisfied. These difficulties facing regression analysis or stochastic modeling provide the motivation for addressing the problem with a fuzzy-rule-based approach.

Concept of Fuzzy-Rule-Based Modeling

In the classical case, a rule is a function formulated with arguments coupled by logical operators, yielding a logical expression and a corresponding response. If the conditions of the rule are fulfilled (the logical expression is true) then the response has to be true. The logical expression is usually formulated with simple uni- and bivariate logical operators.

With a fuzzy rule, binary logic is replaced by fuzzy logic, where a statement and its opposite may both be "true" to different degrees. To classify, say, drought severity by the use of fuzzy logic, each possible class is first described by a set of fuzzy rules. The classification is then done by finding the most appropriate rule for the selected object. This is implemented by calculating for each class the degree of fulfillment (DOF) to which the object fulfills the describing rules. The class with the rule having the highest DOF value, for example "hydrologic drought", is then assigned to the drought. The DOF thus describes the degree of applicability of the rule. This procedure is now formally described, as in Bardossy and Duckstein (1995):

The ith fuzzy rule consists of a set of arguments in the form of K fuzzy sets $(A_{i1}, ..., A_{iK})$ with membership functions and a consequence also in the form of a fuzzy set. For a general input vector of values $(a_1, a_2, ..., a_K)$ the rule is applied as:

$$\text{If} \quad a_1 \text{ is } A_{i,1} \odot a_2 \text{ is } A_{i,2} \odot ... \odot a_K \text{ is } A_{i,K}$$
$$\text{then} \quad (a_1, ..., a_K) \text{ corresponds to } B_i \tag{1}$$

where $\odot$ is any logical operator, specified according to the application. Usually, rules are formulated using AND/OR operators. For example, with K=3, let k=1 correspond to one of the ENSO index classes (e.g., La Niña), k=2 to a relative frequency class of a given CP type (say rare), while k=3 corresponds to a class of spatially averaged geopotential height of the 500 hPa level (say low), and B_1 represents a response such as the "chance for a severe drought". The DOF can be calculated from the membership values $\mu A_k(a_k)$ of inputs, according to the logical operators $\odot$ applied in Equation (1). If this operator is only "AND", then one can take the product of the membership functions:

$$D_i(a_1, ..., a_K) = \mu A_1(a_1) ... \mu A_K(a_K) \tag{2}$$

To sum up, fuzzy-rule-based modeling shows much potential in cases when a causal relationship is well-established but data are scarce and imprecise, and furthermore, when a given input vector may have several contradictory responses which may be true to varying degrees as measured by DOFs of the corresponding fuzzy rule.

This fuzzy-rule-based modeling was applied for flood probabilities under climatic fluctuations (Bogardi et al. 1996).

Application for Drought Index

In our present application, the input variables are the monthly values of SOI and the monthly relative frequencies of CP types; the output variable is the drought index (PMDI). The modeling procedure is as follows:

1. Fuzzy numbers are defined on the input variables. In the present models, we used triangular fuzzy numbers for SOI. The memberships functions defined on SOI values can be seen in Figure 4. The fuzzy representation of CP types is based on the monthly relative frequencies of each CP type. Specifically, triangular fuzzy numbers were defined similarly to SOI, but instead of the SOI values the monthly frequency is used on the x-axis, and the three membership functions are labeled as "rare", "medium", and "frequent". The maximum values of these x axes are the maximum empirical frequencies for each CP type as shown in Figure 5.

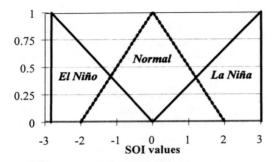

Figure 4. Membership Functions Defined on SOI

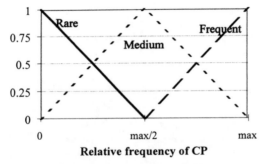

Figure 5. Membership Functions Defined on the Relative
Frequency of a Given CP Type

2. The whole data set of outputs and inputs is divided into two parts: the 33-year training set lasts from 1946 through 1978, and the validation set includes the following 16 years, from 1979 through 1994.

3. Threshold for the drought index is defined. Several thresholds can be tried: all the drought values investigated (case of no threshold), or only dry conditions (months when PMDI was less than 0 or -1).

4. Fuzzy rules are constructed from the training set using various types of algorithms (Bardossy and Duckstein 1995). While the counting algorithm handles each rule with equal importance, in the case of a weighted counting algorithm each rule has a DOF that can be used as weights, a measure of importance. Therefore using weighted technique gives more precise outputs than that without weighting.

In order to provide even better accuracy, in the next steps we defined different DOF thresholds for the events taken into consideration during the rule construction. The threshold value 0.1 seems to result in the best accuracy, better than 0 threshold when every possible event counts, and better than 0.2 when too few events remain to take into account and consequently the number of rules would be too small.

5. Using the constructed rules, crisp values of the drought index are calculated for the validation set with a defuzzification procedure. First, a normed weighted sum combination method was used to get the resultant fuzzy response in the time series. For a given input vector, there can be several rules which are fulfilled to a certain degree, so these DOFs were used as weights. Then crisp responses can be derived with mean defuzzification. The fuzzy mean is the number for which the part of the membership function on its left is in equilibrium with the right side. This equilibrium occurs when the corresponding moments equal. For triangular fuzzy numbers, the fuzzy mean can be calculated easily as the average of the maximum, minimum, and peak values.

6. The constructed rules are verified against the validation set (v). A possible measure of accuracy is:

$$E = \frac{1}{|v|} \sum_{s \in v} \left| R(a_1(s), ..., a_K(s)) - b(s) \right| \tag{3}$$

where $R(a_1(s), ..., a_K(s))$ is the derived response value $(s \in v)$
$b(s)$ is the observed response value $(s \in v)$

Figure 6 presents an example of the preliminary results using the validation set and the corresponding predicted PMDI values. If the dry subset of PMDI is modeled, the measure of accuracy is 1.06. This E-value is better than for a similar model where both dry and wet data are included.

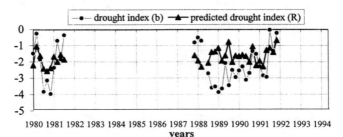

Figure 6. Verification for the Model with DOF = 0.1 for Dry Periods

Conclusions

1. In the state of Nebraska, drought conditions represented by the monthly Palmer index are dependent on global atmospheric circulation patterns and ENSO.
2. Due to the relatively weak statistical dependence, regression analysis or stochastic modeling may have difficulties with quantifying the relationship.
3. Fuzzy logic methods have several advantages in comparison with stochastic techniques. They have relatively simple structures, and they require neither the independency of input data nor large databases.
4. The fuzzy-rule-based approach provided encouraging preliminary results. Specifically, the weighted counting algorithm led to more accurate results than the counting algorithm.

Acknowledgment

Research leading to this paper has been partially supported by a twin grant from the US National Science Foundation to the University of Nebraska (CMS-9614017) and the University of Arizona (CMS-9613654). The additional support of the Center for Infrastructure Research of the University of Nebraska is also acknowledged.

References

Andrade, E.R., and W.D. Sellers. 1988. El Niño and its effects on precipitation in Arizona and western New Mexico. *Journal of Climatology* 8: 403-410.

Bardossy, A., and L. Duckstein. 1995. *Fuzzy-rule-Based Modeling with Applications to Geophysical, Biological, and Engineering Sciences.* Boca Raton, FL: CRC Press.

Bartholy, J., I. Matyasovszky, and I. Bogardi. 1994. Sensitivity of a stochastic spacetime precipitation model on macrocirculation classification. Presentation at European Geophysical Society XIX General Assembly, April 25-29, Grenoble, France.

Bogardi, I., R. Reiter, and H.P. Nachtnebel. 1996. Fuzzy-rule-based estimation of flood probabilities under climatic fluctuations. In Y.Y. Haimes, D. Moser, and E.Z. Stakhiv (Eds.). *Risk-Based Decision Making in Water Resources VII:* 61- 79. New York, NY: American Society of Civil Engineers.

Carlson, R.E., D.P. Todey, and S.E. Taylor. 1996. Midwestern corn yield and weather in relation to extremes of the Southern Oscillation. *Journal of Production Agriculture* 9: 347-352.

Covey, C. 1995. Using paleos to predict future climate: how far can analogy go? *Climate Change* 29: 403-407.

Dickinson, R.E., R.M. Errico, F. Giorgi, and G.T. Bates. 1989. Regional climate model for the western United States. *Climatic Change* 15: 382-422.

Giorgi, F., and L.O. Mearns. 1991. Approaches to the simulation of regional climate change: a review. *Review of Geophysics* 29: 191-216.

Glantz, M.H. (Ed.). 1988. *Societal Responses to Regional Change: Forecasting by Analogy.* Boulder, CO: Westview Press.

Hansen, J., A. Lacis, D. Rind, G. Russel, P. Stone, I. Fung, R. Ruedy, and J. Lerner. 1984. Climate sensitivity: analysis of feedback mechanisms. In *Climate Processes and Climate Sensitivity,* Geophysics Monograph 29: 130-163.

Heddinghause, T.R., and P. Sabol. 1991. A review of the Palmer Drought Severity Index and where do we go from here. In *Proceedings of the Seventh Conference on Applied Climatology:* 242-246.

Horell, J.D., and J.M. Wallace. 1981. Planetary-scale atmospheric phenomena associated with the Southern Oscillation. *Monthly Weather Review* 109: 813-829.

Julian, P.R., and R.M. Chervin. 1978. A study of the Southern Oscillation and Walker Circulation phenomena. *Monthly Weather Review* 106: 1433-1451.

Kiladis, G.N., and H.F. Diaz. 1989. Global climatic anomalies associated with extremes in the Southern Oscillation. *Journal of Climate* 2: 1069-1090.

Matyasovszky, I., I. Bogardi, A. Bardossy, and L. Duckstein. 1993. Estimation of local precipitation statistics reflecting climate change. *Water Resources Research* 29: 3955-3968.

NOAA (National Oceanic and Atmospheric Administration). 1997. SOI time series. http://nic.fb4.noaa.gov:80/data/cddb/cddb/soi

NOAA (National Oceanic and Atmospheric Administration), National Climatic Data Center. 1997. *Modified Palmer Drought Severity Index.* ftp://ftp.ncdc.noaa.gov/pub/data/cirs/9705.pmdi

Pesti, G., B. Shrestha, L. Duckstein, and I. Bogardi. 1996. A fuzzy-rule-based approach to drought assessment. *Water Resources Research* 32: 1741-1747.

Presnell, D. W. 1995. Preferred climate patterns in the United States associated with categorized El Niño/Southern Oscillation (ENSO) events. In *Proceedings of 6th International Meeting on Statistical Climatology,* Galway, Ireland: 587-590.

Rasmusson, E.M., and T.H. Carpenter. 1982. Variations in tropical sea surface temperature and surface wind fields associated with the Southern Oscillation/El Niño. *Monthly Weather Review* 110: 354-384.

Redmond, K.T., and R.W. Koch. 1991. Surface climate and streamflow variability in the western United States and their relationship to large-scale circulation indices. *Water Resources Research* 27: 2381-2399.

Ropelewski, C.F., and M.S. Halpert. 1987. Scale precipitation patterns associated with the El Niño/Southern Oscillation. *Monthly Weather Review* 115: 1606-1626.

Schonher, T., and N. Nicholson. 1989. The relationship between California rainfall and ENSO events. *Journal of Climate* 2: 1258-1269.

Trenberth, K.E., and D.J. Shea. 1987. On the evolution of the Southern Oscillation. *Monthly Weather Review* 115: 3078-3096.

Woolhiser, D.A., and T.O. Keefer. 1993. Southern Oscillation effects on daily precipitation in the southwestern United States. *Water Resources Research* 29: 1287-1295.

Use of Criteria-Based Rankings in Risk-Based
Analysis of Ecosystem Restoration Projects

Charles Yoe[1]

Abstract

Little risk-based analysis is currently being applied to the evaluation of ecosystem restoration projects. Risk-based methods must be cognizant of the fact that there are limited resources to accomplish these investigations. This paper proposes a flexible set of procedures to guide a risk-based analysis that can be adapted to any size study budget. A criteria-based ranking system is then used to identify potentially important types and sources of uncertainty, one of the critical tasks in any risk-based analysis. The procedures and ranking system are applied to the evaluation of a US Corps of Engineers' project using a set of US Fish and Wildlife Service habitat suitability index models. This research demonstrates the feasibility of using a Monte Carlo process to identify the potential range in project outputs once critical input variables have been identified.

Introduction

Environmental policies and programs in the United States have proliferated since the National Economic Policy Act (NEPA) of 1969. Environmental legislation covering historic and cultural resources, fish and wildlife, their habitats, water, air, antiquities, threatened and endangered species, hazardous substances and toxic chemicals, natural resources, landmarks, archaeological sites, anthropological resources, natural processes, wetlands, barrier islands, rivers, bays and other water resources, vast arrays of ecosystems, farmlands, pesticides, food, and other environmental values have been the regular business of not only the US government but of state and local governments as well. In recent years, these initiatives have included a more or less *ad hoc* collection of programs that can be considered environmental investments. These investments include such things as traditional

---

[1]Associate Professor of Economics, College of Notre Dame of Maryland, Economics Department, 4701 N. Charles Street, Baltimore, MD 21210-2476

anti-pollution expenditures and improvements, clean-up of hazardous waste sites, natural resource damage mitigation, fish and wildlife mitigation measures, water and waste treatment, protection of existing natural and environmental resources, and the like. Of particular interest in this paper is a subset of environmental investments called ecosystem restoration projects. Selected federal legislation supporting ecosystem restoration is shown in Table 1.

Table 1. Legislation Supporting Federal Involvement in Ecosystem Restoration

Fish and Wildlife Coordination Act of 1958, as amended
Federal Water Project Recreation Act of 1965, as amended
National Environmental Policy Act of 1969, as amended
Coastal Zone Management Act of 1972, as amended
Water Pollution Control Act of 1972, as amended
Endangered Species Act of 1973, as amended
Water Resources Development Acts of 1986, 1988, 1990, and 1992
Coastal Wetlands Planning, Protection and Restoration Act of 1990 (Title III of PL 101-646)

Ecosystem restoration projects include any management actions taken to recover or re-establish native ecosystems. These projects are typically intended to: 1) restore natural systems to evolutionary environmental conditions, 2) prevent further degradation, or 3) conserve native plants and animals[2]. The US Army Corps of Engineers is a federal agency that has developed an ecosystem restoration program and philosophy (USEPA/USACE 1995) that includes giving budget priority to projects for "restoration of degraded ecosystem functions and values, including its hydrology, plant and animal communities, and/or portions thereof, to a less degraded ecological condition." The Corps' Section 1135 Program[3] provided the opportunity to introduce the use of risk-based analysis to the estimation of ecosystem restoration as part of a developmental research program funded by the Institute of Water Resources under its Evaluation of Environmental Investments Research Program (EEIRP).

The planning, design, construction or implementation, operation, and maintenance of ecosystem restoration projects are new ventures. In truth, many of

---

[2]This working definition was obtained from an Internet discussion paper citing the author simply as Bill Pell, Ecologist. We regret we are unable to provide a more complete citation of the source of these thoughts.

[3]Section 1135 of WRDA 86, PL 99-662, as amended by Section 41 of WRDA 88, PL 100-676, Section 304 of WRDA 90, PL 101-640, and Section 202 of WRDA 92, PL 102-580 authorizes the Secretary of the Army to carry out a program to make modifications in the structures and operations of water resources projects supervised by the secretary that are feasible and consistent with the authorized project purposes, and which would improve the quality of the environment in the public interest.

the projects are science-based best guesses about how to achieve the goals of an ecosystem restoration project. There is considerable uncertainty attending the entire program. The general sources of risk and uncertainty in ecosystem restoration planning have been identified in IWR Report 96-R-8 (USACE 1996) *An Introduction to Risk and Uncertainty in the Evaluation of Environmental Investments*. They tend to parallel the steps of the planning process: problem identification, resource inventories and forecasts, plan formulation, evaluation, comparison, and selection of plans. One of the most significant uncertainties to emerge from a typical planning process, however, concerns project performance.

A Corps planner, describing ecosystem-restoration management measures remarked, "We're fiddling with the landscape, the spacing of plants, height of fill, number of duck boxes, and we're using our hydraulics and hydrology and water quality data to manipulate water. Will this result in what we want? If we build it, will they come?" This frank commentary reveals the truth that planners may not always know how well their projects will succeed in achieving planned objectives. Such a significant uncertainty must be addressed in the evaluation of ecosystem restoration projects.

Starting with the next section, the remainder of this paper uses a case study from the Corps' Section 1135 Program to introduce risk-based analysis to the evaluation of ecosystem restoration projects. Because the resources available to complete this kind of evaluation can vary from meager to substantial, planners need a flexible set of procedures to guide their risk-based analysis. Such a set of procedures is described in Section 3.

One of the more difficult tasks in conducting a risk-based analysis in such a new field is deciding which of the many uncertainties are important enough to address and which can be safely overlooked. Section 4 provides a criteria-based method for ranking and deciding on the uncertainties needing to be addressed in a typical habitat evaluation procedure. The results of the case study are presented in Section 5 and some conclusions follow in the final section of this paper.

Case Study

The Brown Sugar River[4] once supported a warm-water fishery. In the 1950s the Tentshow Dam was constructed to provide hydroelectric power. The warm-water fishery was adversely impacted by the cold, low dissolved oxygen (DO) hypolimnetic water released from the dam for the generation of electricity. Re-establishment of the warm-water fishery was not considered feasible and efforts began to introduce a cold-water rainbow trout fishery to the river. Efforts to establish a year-round cold-water fishery have been thwarted by the low DO and low flows. Figure 1 provides a stylized map of the project area.

---

[4]Fictional names are used to protect the identity of the project.

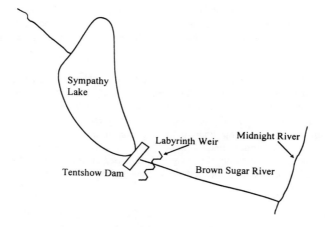

Figure 1. Map of Project Area

There is little or no growth in the trout fishery and the river remains a put-and-take fishery despite a reservoir release program. The proposed restoration measures[5] consist of a 2,100-foot zigzag labyrinth-shaped weir across 242 feet of river, 2,000 feet downstream of the dam. The weir crest would be 3.5 feet above normal water surface and water would flow over the weir at a depth of about 6 inches, for a head differential of 4 feet. With pipes strategically located to facilitate low flow releases from the weir, the project is expected to address both the DO and low flow problems.

The cost of the structure would be about $3.35 million with annual operation and maintenance costs of $1,000. The primary output of this project would be a more viable cold-water fishery. Although there were eight variations of this plan, this paper presents the results for a single alternative.

Habitat Evaluation Procedures

How are the outputs of a labyrinth-shaped weir measured in a situation such as this? In an ecosystem restoration project the primary output is often improvement in ecosystem function (Schroeder and Haire 1993). Habitat is one ecological

---

[5]It is interesting to note that this project does not seem to meet any of the definitions of ecosystem restoration offered at the beginning of this paper. The cold-water fish this project is trying to establish is a non-indigenous species. This hardly seems to be ecosystem restoration. Nonetheless, this is the case study we were asked to investigate and the principles and findings of this analysis are in no way reduced.

resource commonly used to represent ecosystem function. In general, more or better habitats are assumed to indicate better ecosystem function. Thus, habitat improvements are commonly used as surrogate measures of ecosystem restoration-project outputs.

Because there can be many different kinds of habitats in an ecosystem, a problem arises in describing habitat improvements. How do we describe such complex concepts in a compact yet serviceable way? Although many options are available, it is common practice to identify a few key species from an ecosystem and discuss the changes in their habitats. The presumption is that if the species are carefully chosen in a representative manner this can reasonably serve as an indicator of overall ecosystem function. For example, if a species at the top of the food chain is doing well, chances are good that the species below it in the food chain are also doing well.

Changes in the habitats of these indicator species are frequently measured in habitat units for the more common and less complex ecosystem restoration studies. A habitat unit (HU) is a theoretical indicator that combines the quantity and quality dimensions of a habitat in a simple mathematical way.

Habitat quantity is estimated as some physical quantity of terrestrial or aquatic habitat, usually acres. *Habitat quality*, however, is quantified via an index number between zero and one. An index of 1 indicates the habitat in question is optimal for the specific indicator species under consideration. An index of zero indicates unsuitable habitat. Intermediate values indicate the relative suitability of the habitat.

Changes in habitat units can be used to represent the ecological impacts of habitat management activities or planned improvements. For example, suppose we have 10 acres of land with an average suitability index of 0.6. Such land would yield 6 habitat units:

$$10 \text{ acres} \times 0.6 = 6.0 \text{ habitat units} \tag{1}$$

Now suppose an ecosystem restoration plan would double the acres of habitat and increase their quality from 0.6 to 0.8. The result would be 16 habitat units:

$$20 \text{ acres} \times 0.8 = 16 \text{ habitat units} \tag{2}$$

The plan would result in a net output of 10 additional habitat units. The increase of 10 habitat units is used to represent an improvement in overall ecosystem function.

There are many methods for estimating ecosystem function improvements in this general way. The habitat evaluation procedures (HEP) methodology of the US

Fish and Wildlife Service (1980) was used in the case study because it is believed to be the methodology in widest use in ecological restoration studies at this time.

The HEP analysis index number is called the habitat suitability index (HSI). The estimation of habitat units in an HEP analysis can be formally defined as:

$$\text{Quantity} \times \text{Quality} = \text{Habitat Units} \qquad (3a)$$
$$\text{Acres} \times \text{Habitat Suitability Index} = \text{Habitat Units} \qquad (3b)$$

The specific methodology for estimating an HSI is typically described in a habitat suitability model. This model reviews the scientific literature pertaining to the species of interest to identify a set of habitat variables. These variables describe those environmental factors that are important to the survival, growth, and reproduction of the species. For the rainbow trout, 18 habitat variables were identified as shown in Figure 2.

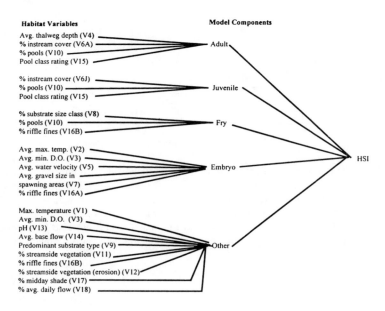

Figure 2. Rainbow Trout HSI Model

The suitability of a given habitat is evaluated in terms of each of the relevant habitat variables by means of a suitability index (SI). A suitability index graph for the rainbow trout is shown in Figure 3. The curves are built on the assumption that increments of the habitat variable plotted on the x-axis can be directly converted into an index of suitability from 0.0 to 1.0 for the species. Thus, the SI index number is at best the model authors' science-based subjective judgment of the relationship between the measurement of a variable and its suitability for the species of interest.

The example in Figure 3 shows that when the percent of substrate in the 10-40 cm size group is zero, the habitat is lacking in escape cover for fry and small juveniles. Unlike an HSI of zero, an SI of zero need not imply that the habitat is totally unsuitable for the species. The overall suitability of the habitat as indicated by the HSI reflects a composite trade-off of the relative strengths and weaknesses of the habitat's various characteristics.

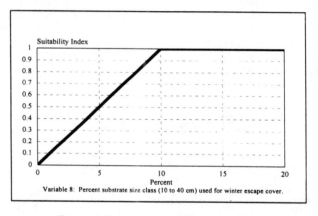

Figure 3. Substrate Suitability Index Graph

The diverse habitat variables are grouped into "components" or "life requisites." For example, the rainbow trout has the following components: fry (C_F), embryo (C_E), juvenile (C_J), adult (C_A), food (C_{OF}), and water quality (C_{OQ}). These components are mathematical combinations of the SIs for the different habitat variables that define each component. Component values are also index numbers between zero and 1. For example, the food component for trout is defined as follows:

$$C_{OF} = \frac{(V_9 x V_{16})^5 + V_{11}}{2} \qquad (4)$$

where V_i is the SI for the ith habitat variable. The model components are then combined mathematically to produce an overall HSI. Thus, the general progression of an HSI model is:

Habitat variable measurements => Suitability indices =>   (5)
Component indices => Habitat suitability index

Table 2. Eight Steps to Risk Analysis

1. Select Analytic Framework for Environmental Outputs
   a. Review and select models/techniques for evaluating project outputs
   b. Understand the models you use
   c. Make an informed choice of tools

2. Identify Types and Sources of Uncertainty
   a. Know the types of uncertainty
   b. Know the sources of uncertainty

3. Identify Potential Key Variables
   a. Determine potential importance

4. Design Risk Analysis
   a. Assess importance of analysis
   b. Review tools available
   c. Select tools

5. Collect Data
   a. Consider data needs of risk analysis
   b. Define your terminology
   c. Design a data collection methodology
   d. Use interval estimates
   e. Use distributions

6. Identify Major Uncertainties
   a. Review the potential key variables and identify actual key variables
   b. Describe key uncertainties
   c. Pay attention to key sources of uncertainty

7. Do Risk-Based Analysis
   a. Do the analysis
   b. Verify your analysis
   c. Meet or exceed minimum expectations of risk analysis
   d. Document your analysis

8. Communicate Results of Risk Analysis
   b. Tell the risk analysis story
   c. Meet or exceed minimum reporting requirements
   d. Serve the risk management function

Indicator species for the case study included the rainbow trout, the channel catfish, and the smallmouth bass.

Procedures for Risk-Based Analysis of Ecosystem Restoration Outputs

The Corps' EEIRP program was unable to discover any instances in which risk-based analysis was being applied in ecosystem restoration projects to improve the decision-making process. Development of a flexible set of procedures was one of the first steps in this analysis. Planners need guidance that provides enough structure to understand what is required but is flexible enough to allow a study team to adjust the analysis to the resources and decision requirements of an investigation.

Table 2 presents a summary of an eight-step set of procedures that strikes a balance between structure and flexibility that should meet the needs of most analysts who are beginning to do risk-based analyses of project outputs. A detailed discussion of this procedure can be found in the IWR Report 97-R-7, *Risk and Uncertainty Analysis Procedures for the Evaluation of Environmental Outputs* (Yoe and Skaggs 1997). Of particular interest is the requirement to identify potential key sources of uncertainty in preparation for performing a risk-based analysis (Steps 3 and 6). As the above brief description of the HEP method suggests, uncertainty begins with the structure of the HSI model and the analyst's choice of the indicator species. Given the choice of rainbow trout and acceptance of the USFWS HSI model (McMahon and Terrell 1982; Raleigh et al. 1984; Stuber et al. 1982), the primary sources of uncertainty become the values of the habitat variables. In a low-budget, short-duration study with three indicator species, each with over a dozen habitat variables, it is critically important for the analyst to understand which of these dozens of variables can have the most potential impact on habitat units and on the ultimate decision to pursue a course of action or not. An inexpensive and effective process for identifying potential key uncertainties is described below.

Criteria-Based Ranking of Uncertainties

This section offers a generic process to assist analysts in the identification of potentially important habitat variables in a risk analysis. The method, called criteria-based ranking, is useful when the important variables aren't obvious or there are so many of them they cannot all be addressed. The value of the technique is that it allows the analyst to identify a small set of tailor-made criteria that can be used to organize information and place potentially important uncertainties in some order of priority. The method is described in the seven steps that follow.

1. Criteria

The first step is to identify the criteria used to rank the potentially uncertain variables. The criteria can vary from project to project. Criteria should be designed to reflect the most important uncertainties in a given situation. Some sample criteria for selecting habitat variables could include:

1. Can it cause the HSI to go to zero?
2. Does it have an SI of zero?
3. Can it be directly affected by alternative plans?
4. Can it be indirectly affected by alternative plans?
5. Does anyone say it is important?
6. Can the variable impact any charismatic species?
7. Can the variable impact any threatened or endangered species?

Criteria-based ranking works best when the number of criteria used is limited. Generally, it would be desirable to keep the number of criteria to a maximum of three or four for this screening technique to be effectively applied with pencil and paper.

Once a criterion is chosen, a variable number of scenarios (usually three) are defined. The criteria as well as the scenario descriptions are site- and study-specific. They are based on the professional opinions of the study analysts; hence, they are subjective by nature. An example of some criteria and their scenarios for the case study follows. High, medium, and low refer to the potential importance of the scenario for its criterion.

Criterion 1. Habitat variable, can cause HSI to go to zero.

High: If SI for variable is zero, HSI will be zero.
Medium: If SI for variable is zero, HSI will be low.
Low: HSI is determined by other variables.

Criterion 2. Others say the habitat variable is important.

High: Published studies and/or the non-federal partner
           identify the variable as important.
Medium: Stakeholders say the variable is important.
Low: Neither published reports nor stakeholders indicate
           that the variables are important.

Criterion 3. Alternative plans can affect the habitat variable.

High: One or more potential alternative plans directly affects the
           variable.
Medium: One or more potential alternative plans indirectly
           affects the variable.
Low: The variable is not affected by a potential alternative plan.

Ideally, the scenarios would be inclusive of all possible states of the world. This will rarely be feasible. To do so would require far too many scenarios. Bearing in mind that this is a screening tool, it is usually more practical to define three relatively general scenarios and then fit each case into one of these scenarios. If it appears that doing so could result in egregious error, then add another scenario.

The method is easiest when all the criteria are considered of equal importance. If this is neither practical nor realistic, then the weighting scheme should be defined at this step. For example, we might say Criterion 1 is twice as important as Criterion 2 and three times as important as Criterion 3. These weights would be reflected in the ratings described below.

2. Ratings

In this step, the study team critically evaluates the available information and uses subjective expert judgment to rate each variable. The rating means each habitat variable is assigned a most-likely scenario. For example, DO and temperature would be assigned to the high-importance scenario under Criterion 2 because of previous reports by other federal agencies. A sample rating, using selected habitat variables of the trout model, is shown in Table 3.

Table 3. Sample Criteria-Based Rating for Rainbow Trout

Habitat Variable	Criteria 1	Criteria 2	Criteria 3
V_1 Maximum temperature	H	H	H
V_3 Minimum DO	H	H	H
V_4 Thalweg depth	H	L	L
V_6 Percent cover	H	L	L
V_9 Substrate class	M	L	L
V_{10} Percent pools	H	L	L
V_{11} Percent riparian vegetation	M	L	L
V_{12} Percent ground cover	H	L	L
V_{13} pH	H	L	M
V_{14} Average annual base flow	H	M	H
V_{15} Pool class	H	L	L
V_{16} Percent fines	M	L	L

3. Possible Combinations

In this step all the possible combinations of scenario ratings for the criteria are listed in descending order of possible importance. This requires analysts to pay special attention when the criteria are not weighed equally. A sample listing of all possible combinations with equally weighted criteria is shown in Table 4.

Table 4. Possible Combinations for Rainbow Trout

HHH	Greatest potential importance
HHM, HMH, MHH	High potential importance
HHL, HLH, LHH, HMM, MMH, MHM	Above-average potential imp.
HLM, MHL, HML, LMH, MLH, MMM, LHM	Moderate potential importance
HLL, LHL, LLH, MML, LMM, MLM	Below-average potential imp.
MLL, LML, LLM	Low potential importance
LLL	Least potential importance

The table reveals the subjectivity of the method. The value of the technique is that it provides analysts with an organized and consistent approach for whittling a long list of potentially important variables down to those on which they will focus their attention. The criteria and scenarios developed in Step 1 make the analysts' subjective judgments transparent to others. If anyone disagrees with the criteria or scenarios, they are free to modify the technique and apply it themselves.

4. Rank

The habitat variables are ranked according to descending relative importance in subjective clusters. This combines Steps 2 and 3. The rankings for the rainbow trout are provided in Table 5.

There is uncertainty attending estimates of each variable. The criteria-based ranking procedure has enabled us to define our own criteria and scenarios and to separate the 12 different habitat variables used into 5 subjective groupings. The analysts must now decide which, if any, of these groupings they should address. Any variable that presents a "high potential importance" or greater should be measured carefully. This means temperature (V_1), DO (V_3), and flow (V_{14}) warrant close scrutiny in the case study. Less emphasis would be placed on the other variables. A similar process would be followed for each of the other HSI models used.

Table 5. Criteria-Based Ranking for Rainbow Trout

Habitat Variable	Rating	Ranking
V_1, V_3	HHH	Greatest potential importance
V_{14}	HMH	High potential importance
V_{13}	HLM	Moderate potential imp.
$V_4, V_6, V_{10}, V_{12}, V_{15}$	HLL	Below-average potential imp.
V_9, V_{11}, V_{16}	MLL	Low potential importance

5. Add Criteria

The next step is for analysts to use their expert judgment to assess the accuracy of the uncertainty-importance ranking that resulted from the initial set of criteria. For argument's sake, suppose the analysts all thought that pH (V_{13}) should have come out as the single most important factor in this analysis. Instead, it came out fourth. It would be very difficult to justify focusing solely on pH based on this analysis, as that would require leapfrogging over more important variables.

If the analysts believe a variable is ranked too low, it could be because the original criteria did not address some dimension of importance. In that case, it may be appropriate to add another criterion which should address that missing dimension. For example, it's perfectly permissible to add a criterion that would advance pH up the risk ranking as long as you describe what you did and why you did it. Although we do not take that step here, the following steps describe how to proceed once this is done.

6. New Combined Rating

In this step, the habitat variables would be rated again, this time against four criteria — the original three and the new one. The new combined ratings, from HHHH to LLLL, would presumably result in a change in the ranking of the potential-importance-of-the-habitat variables, otherwise there would have been little reason to add a criterion.

7. New Ranking

In order to provide a new ranking, a fresh set of possible combinations must first be developed. When all the combinations of the three scenarios for four criteria are ranked, it becomes clear why this process works best for a limited number of criteria. There is no reason why a large number of criteria cannot be used if the technique is built into a spreadsheet environment or is used with some multi-criteria decision analysis software such as, for example, Expert Choice[6]. Criteria-based ranking is presented here as a simple tool that can be done with pencil, paper, and a careful thought process. Once the new table of possible combinations is created, the habitat variables would be ranked again as was done in Table 4.

The value of this method of identifying potentially important variables in a risk-based analysis is that it provides a transparent approach for reducing a long list of variables you need to measure to a short list whose measurement could significantly affect the decision to be made as a result of your analysis.

---

[6]Advanced decision-support software made by Expert Choice, Inc. of Pittsburgh, PA.

Results of Case Study

Once the potentially important sources of uncertainty in the estimation of project outputs have been identified, the uncertainty must be addressed. This means, at a minimum, that greater care must be taken in measuring these habitat variables than in measuring others. The uncertainty present in this measurement needs to be addressed in a manner consistent with the design of the risk analysis (see Table 2). Generally, this would mean at least conducting some sensitivity analysis. A Monte Carlo simulation was used in the case study to investigate the impact of the uncertainty of key variables on project outputs. A Monte Carlo procedure requires using a probability distribution to describe the potential uncertainty.

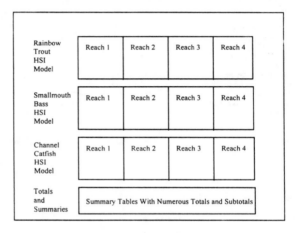

Figure 4. Model Architecture

The overall architecture of the simulation model is shown in Figure 4. Each reach of each HSI model contains a component similar to that shown in Figure 5; hence, there are 12 such model components. Habitat variable measurements are entered in columns B (Without Project condition) and E (With Project condition). The measurements have been entered as probability distributions and the change in habitat variables reported in cell D41 was calculated 10,000 times (for each of the 12 components) to describe the range of potential benefits from the proposed labyrinth weir.

The results of the analysis of all 12 model components are summarized in Figure 6. The observed range in potential project outputs was 66 HUs, from a minimum of 29.4 HUs to a maximum of 95.4 HUs. The weir project is expected to produce an output of about 57.6 additional HUs. Although this range in HUs is relatively small, it does imply that a small absolute variation in output can mean a large relative change. The costs per unit of output, using the $3.35 million cost

	A	B	C	D	E	F	G
1	TROUT						
2		Site 2			Site 2		
3		Without Project Condition			With Project Condition		
4	**Habitat Variable**	Measure	SI		Measure	SI	
5	V1: Maximum Temperature						
6	A=resident rainbow trout	25.8	0		24.2	0.174026	
7	V2: Maximum Temperature (embryo)						
8	V3: Minimum dissolved oxygen	Use this==	0		Use this==	0.567055	
9	A=<=15 Degrees C	0	0		0	0	
10	B=>15 Degrees C	1	0		6.533333	0.567055	
11	V4: Average Thalweg Depth	Use this==	1		Use this==	1	
12	Average Stream Width	6			6		
13	A = <= 5m stream width	68.06667	1		68.06667	1	
14	B= > 5 m stream width	68.06667	1		68.06667	1	
15	V5: Average Velocity						
16	V6: % Cover						
17	V6: % Cover, A = adults	8.5	0.616072		8.5	0.616072	
18	V7: Substrate Size						
19	V8: % Substrate Size						
20	V9: Substrate Class (food)	1	1		1	1	
21	V10: % pools	68.33333	0.968844		68.33333	0.968844	
22	V11: % riparian vegetation	104.8333	0.745054		104.8333	0.745054	
23	V12: % ground cover (erosion)	87.5	1		87.5	1	
24	V13: Maximum-minimum PH	7.5	1		7.5	1	
25	V14: Average annual base flow	3	0.06		3	0.06	
26	V15: Pool class	3	0.3		3	0.3	
27	V16: % fines						
28	B = riffle-run	4	1		4	1	
29	V17: % shade	3.833333	0.353667		3.833333	0.353667	
30	V18: % average daily flow						
31							
32	Requisites:	With	Without	Change			
33	Adult (CA)						
34	(V10*V15)^0.5=	0.539123	0.539123	0			
35	Is V6 > (V10*V15)^0.5 ? (1=yes, 0=no)	1	1	0			
36	Choose CA Equation	0.332139	0.332139	0			
37	Adult (CA)	0.332139	0.332139	0			
38	Other (CO)	0	0.521828	0.521828			
39	HSI	0	0.416316	0.416316			
40	Area in Acres	22.52	22.52	0			
41	Habitat Units	0	9.375445	9.375445			

Figure 5. Sample HSI Spreadsheet Model

estimate, could vary from a low of $35.1 thousand to a high of $113.9 thousand per HU. The expected cost of an HU from this particular plan is $58.2 thousand.

Conclusions

Risk analysis for its own sake has no place in ecosystem restoration studies. If it is to be done, it must be inexpensive and straightforward and it must inform the decision process. If risk analysis procedures are to be helpful in environmental investment decisions, they must be flexible and adaptable to the needs of many different types of ecosystem restoration studies. Such a set of procedures, in eight steps, has been introduced here.

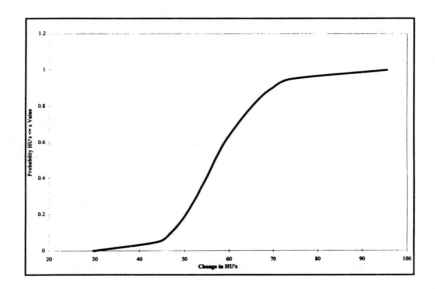

Figure 6. CDF for Weir Outputs

There are few people with any experience using risk-based analysis in ecosystem restoration projects. In a very real sense, this means analysts will be creating the tools and methods as they proceed. To aid in the introduction of risk-based analysis to this type of project, evaluation field analysts need clear guidance and useful tools. One of the most important questions every analyst will face sometimes is where to focus attention when the data requirements for habitat evaluation models overwhelm the resources available for these analyses.

Perhaps the major contribution of this paper has been to offer a transparent (meaning "this is what we did and this is why we did it") process for identifying those habitat variables that deserve the most attention in an analysis. This has been done by reviving and formalizing a criteria-based ranking system, a qualitative risk assessment tool that first appeared several decades ago but has been largely forgotten in favor of more sophisticated quantitative risk-assessment tools. This simple technique proves itself a valuable implement in the risk assessor's toolbox. It also dovetails nicely with the use of sensitivity analysis or Monte Carlo simulation. Experimentation with the procedures offered here and other approaches to risk analysis in ecosystem restoration is a prime objective for future research in this field.

Acknowledgments

This research was supported in part by the Institute for Water Resources (IWR) under an Evaluation of Environmental Investments Program contract let to the Greeley-Polhemus Group, Inc. for whom the author works.. The author gratefully acknowledges the contributions of Mr. Leigh Skaggs of IWR to the work that contributed to this paper.

References

McMahon, T.E., and J.W. Terrell. 1982. *Habitat Suitability Index Models: Channel Catfish.* FWS/OBS-82/10.2, Fish and Wildlife Service, US Department of Interior, Washington, DC.

Raleigh, R.F., T. Hickman, R.C. Solomon, and P.C. Nelson. 1984. *Habitat Suitability Index Models: Rainbow Trout.* FWS/OBS-82/10.60, Fish and Wildlife Service, US Department of Interior, Washington, DC.

Schroeder, R.L., and S.L. Haire. 1993. *Guidelines for the Development of Community-Level Habitat Evaluation Models.* Fish and Wildlife Service, US Department of the Interior, Washington, DC.

Stuber, R.J., G. Gebhart, and O. E. Maughan. 1982. *Habitat Suitability Index Models: Largemouth Bass.* FWS/OBS-82/10.16, Fish and Wildlife Service, US Department of Interior, Washington, DC.

USACE. 1996. *An Introduction to Risk and Uncertainty in the Evaluation of Environmental Investments.* IWR Report 96-R-8, Institute for Water Resources, US Army Corps of Engineers, Alexandria, VA.

USEPA, and USACE. 1995. *Ecosystem Restoration in the Civil Works Program, Appendix C.* EC 1105-2-210, US Environmental Protection Agency and US Army Corps of Engineers, Washington, DC.

US Fish and Wildlife Service. 1980. *Habitat Evaluation Procedures (HEP).* ESMOZ, Department of the Interior, Washington, DC.

Yoe, C.E., and L. Skaggs. 1997. *Risk and Uncertainty Analysis Procedures for the Evaluation of Environmental Outputs.* IWR Report 97-R-7, US Army Corps of Engineers, Alexandria, VA.

Ecological Risk Assessment

James K. Andreasen[1] and Susan Braen Norton[2]

Abstract

Ecological risk assessment, which grew out of the need to quantitatively evaluate the effects of human activities on non-human components of the environment, provides a critical element for environmental decision making by giving risk managers a process for considering available scientific information in selecting a course of action. Ecological risk assessment organizes and analyzes data, information, assumptions, and uncertainties to evaluate the likelihood of adverse ecological effects.

Introduction and Overview of the Ecological Risk Assessment Process

Ecological risk assessment "evaluates the likelihood that adverse ecological effects may occur or are occurring as a result of exposure to one or more stressors" (US Environmental Protection Agency 1992). It is a process for organizing and analyzing data, information, assumptions, and uncertainties to evaluate the likelihood of such adverse effects. Ecological risk assessment grew out of the need to quantitatively evaluate the effects of human activities on non-human components of the environment. It provides a critical element for environmental decision making by giving risk managers a process for considering available scientific information along with other factors (e.g., social, legal, political, or economic) in selecting a course of action.

[1]Ecologist, US Environmental Protection Agency, Office of Research and Development, National Center for Environmental Assessment (8623), 401 M Street SW, Washington, DC 20460
[2]Ecologist, US Environmental Protection Agency, Office of Research and Development, National Center for Environmental Assessment (8623), 401 M Street SW, Washington, DC 20460

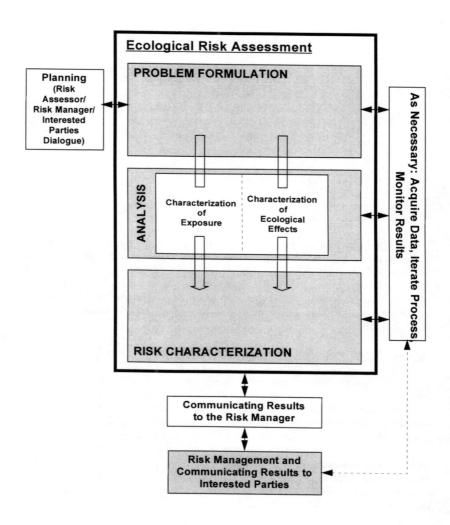

Figure 1. Framework for Ecological Risk Assessment

The risk assessment process is illustrated in Figure 1. At the outset, the views of managers and other interested parties are critical for ensuring that the results of the assessment can be used to support a management decision (Moore and Biddinger 1995). These views could include articulating management goals and alternatives and determining the resources available to conduct the assessment. In regulatory programs, the planning step usually begins by reviewing applicable regulatory mandates (USEPA 1997a). For example, the Clean Air Act requires that EPA set standards for air pollutants that may reasonably be anticipated to endanger public health or welfare, where welfare includes effects on soils, water, crops, vegetation, animals, wildlife, weather, visibility, and climate (CAA § 109). Assessment usually relies on existing data, and the assessment process will often uncover data gaps which need to be filled.

Figure 1 shows several processes that, although outside of the risk assessment process itself, are essential for its successful completion. The box on the right-hand side of Figure 1 represents data collection and monitoring. Data form the basis of risk analysis; these data may have been collected specifically for a risk assessment or developed for another purpose. In the latter case, assessors need to critically evaluate the data to ensure that they can support the assessment. Risk assessment results can provide important input to monitoring plans used to evaluate the efficacy of a risk management decision. For example, if the decision is to mitigate risks through exposure reduction, monitoring can help determine whether the desired reductions in exposure (and effects) have been achieved. Monitoring is also necessary to determine the extent and nature of any ecological recovery that may occur and whether adjustments to the management approach are needed (Holling 1978). Finally, experience obtained by using focused monitoring results to evaluate risk assessment predictions can help improve the risk assessment process.

Both risk managers and risk assessors bring valuable perspectives to the initial planning activities for an ecological risk assessment. Risk managers, who can come from diverse backgrounds including business, law, and engineering, are charged with protecting environmental values. They can ensure that the risk assessment will provide information relevant to making a decision. Ecological risk assessors, usually biologists or ecologists, ensure that science is effectively used to address ecological concerns. Both assessors and managers evaluate the potential value of conducting a risk assessment to address identified problems. Further objectives of the initial planning process are to establish management goals that are agreed upon, are clearly articulated, and contain a way to measure success. Additional goals are to determine the purpose for the risk assessment by defining the decisions to be made within the context of the management goals, and to agree upon the scope, complexity, and focus of the risk assessment, including the expected output and available resources.

As discussed in the proposed Ecological Risk Guidelines (USEPA 1996), an ecological risk assessment consists of three phases: problem formulation, analysis, and risk characterization. Included within problem formulation are identifying goals and assessment endpoints, preparing a conceptual model, and developing an analysis plan. The analysis phase involves evaluating exposure to stressors and the relationship between stressor levels and ecological effects. Risk characterization provides risk estimates by integrating exposure and stressor-response profiles, describing risks by discussing lines of evidence and determining ecological adversity, and summarizing the finding in a report. The interface and dialogue between risk assessors and risk managers at the beginning and end of the risk assessment is critical for ensuring that the results can be used to support a management decision. The remainder of this paper discusses the three phases of ecological risk assessment in greater detail.

Problem Formulation

Problem formulation is critical to the process and provides a foundation upon which the entire risk assessment depends. Successful completion of this phase depends on the quality of three products: assessment endpoints, conceptual models, and an analysis plan. The basis for problem formulation is the integration of available information on the sources of stressors and stressor characteristics, exposure, the ecosystem(s) potentially at risk, and the ecological effects of those stressors. Often, complete information is not available at the beginning of a risk assessment. When this is the case, the process of problem formulation helps to identify where key data are missing and provides the framework for further research to gather additional data. When information of the appropriate type and of sufficient quality and quantity exists, problem formulation can proceed effectively. If new data cannot be collected, the risk assessment will depend on existing data and what can be extrapolated from it. Where little data exist, clearly articulating the limitations of conclusions or uncertainty from the risk assessment becomes increasingly critical in risk characterization. Since problem formulation is inherently interactive and iterative, not linear, substantial reevaluation is expected to occur within and among all the products of the problem formulation.

Assessment endpoints are "explicit expressions of the actual environmental value that is to be protected" (US Environmental Protection Agency 1992) that link the risk assessment to management concerns. A critical component of problem formulation is developing the link between measurable endpoints (e.g., amount of pesticide to use to protect a crop) and the management goal (e.g., preventing unacceptable residues on the final products). This link is aided by defining assessment endpoints which are intermediate between broad management goals and specific measurements. The assessment endpoint defines the specific subject of protection (the entity) and that entity's characteristics of interest (its attributes) (Suter 1990; USEPA 1997b). For example, an assessment endpoint may be the yield of a particular crop, or the growth and survival of trees. With the objectives and assessment endpoints identified, appropriate ecological measurements can be chosen. Effective assessment endpoints are ecologically relevant to the ecosystem they represent and susceptible to the stressors of concern. In addition, assessment endpoints that represent societal values and management goals increase the likelihood that the risk assessment will be used in management decisions. Assessment endpoints that fulfill all three criteria provide the best foundation for an effective risk assessment.

Potential interactions between assessment endpoints and stressors are explored by developing a conceptual model. Conceptual models link anthropic activities with stressors and evaluate interrelationships between exposure pathways, ecological effects, and ecological receptors. Conceptual models include two principal components: risk hypotheses and a conceptual model diagram.

Risk hypotheses describe predicted relationships between stressor, exposure, and assessment endpoint response. Risk hypotheses are hypotheses in the broad scientific sense; they do not necessarily involve statistical testing of null and alternative hypotheses or any particular analytical approach. Risk hypotheses may predict the effects of a stressor (e.g., a chemical release) or they may postulate which stressors may have caused observed ecological effects. Key risk hypotheses are identified for subsequent evaluation in the risk assessment.

A good way to express the relationships described by risk hypotheses is through a *conceptual model diagram*. This is a useful tool for communicating important pathways in a clear and concise way and for identifying major sources of uncertainty. These diagrams and risk hypotheses are used to identify the most important pathways and relationships that will be evaluated in the analysis phase. In an analysis plan, risk assessors justify what will be done in the assessment as well as what will not be done. The analysis plan also describes the data and measures to be used and how risks will be characterized.

Analysis

The analysis phase, which follows problem formulation, includes two principal activities: characterization of exposure and characterization of ecological effects. The conceptual model is used to guide the technical evaluation of data to reach conclusions about ecological exposure and the relationships between stressors and ecological effects. In the characterization of ecological exposures, data are used to describe sources of stressors, the distribution of the stressors in the environment, and the contact or co-occurrence of stressors with ecological receptors.

The products of the analysis phase are summary profiles that describe exposure and the stressor-response relationships. These profiles may be written documents or modules of a larger process model. The exposure profile identifies receptors and exposure pathways and describes the intensity as well as the spatial and temporal extent of exposure. The exposure profile also describes the impacts of variability and uncertainty on exposure estimates and reaches a conclusion about the likelihood that exposure will occur.

The stressor-response profile may evaluate single species, populations, general trophic levels, communities, ecosystems, or landscapes — whatever is appropriate for the assessment endpoints. For example, if a single species is affected, appropriate parameters should be represented, such as effects on mortality, growth, and reproduction, while at the community level, effects may be summarized in terms of structure or function depending on the assessment endpoint. The stressor-response profile summarizes the nature and intensity of effect(s), the time scale for recovery (where appropriate), causal information linking the stressor with observed effects, and uncertainties associated with the analysis.

Risk Characterization

The final phase of an ecological risk assessment is risk characterization. During this phase, risks are estimated and interpreted and the strengths, limitations, assumptions, and major uncertainties are summarized. Risks are estimated by integrating exposure and stressor-response profiles using a wide range of techniques. These include comparisons of point estimates or distributions of exposure and effects data, process models, and empirical approaches such as field observational data. In risk characterization, the information on exposure and stressor-response relationships is integrated to estimate risk. All of the evidence is brought together to reach final conclusions about the likelihood and consequences of effects. Ecological risk assessors describe risks by evaluating the evidence supporting or refuting the risk estimate(s) and interpreting the adverse effects on the assessment endpoint. Criteria for evaluating adversity include the nature and intensity of effects, spatial and temporal scales, and the potential for recovery. Agreement among different lines of evidence of risk increases confidence in the conclusions of a risk assessment. Effective risk characterizations acknowledge uncertainties and assumptions, and separate scientific conclusions from policy judgments (USEPA 1995).

When risk characterization is complete, a report describing the risk assessment can be prepared. The report may be relatively brief or extensive, depending on the nature of the assessment, the resources available for it, and the information required to support a risk management decision. Report elements may include:

- a description of risk assessor/risk manager planning results,
- a review of the conceptual model and the assessment endpoints,
- a discussion of the major data sources and analytical procedures used,
- a review of the stressor-response and exposure profiles,
- a description of risks to the assessment endpoints, including risk estimates

and adversity evaluations,
- a summary of major areas of uncertainty and the approaches used to address them, and
- a discussion of science policy judgments or default assumptions used to bridge information gaps, and the basis for these assumptions.

Discussion

After the risk assessment is completed, risk managers may consider whether additional follow-up activities are required. Depending on the importance of the assessment, the confidence level in the assessment results, and available resources, it may be advisable to conduct another iteration of the risk assessment in order to facilitate a final management decision. Ecological risk assessments are frequently designed in sequential tiers that proceed from simple, relatively inexpensive evaluations to more costly and complex assessments. Initial tiers are based on conservative assumptions, such as maximum exposure and ecological sensitivity. When an early tier cannot sufficiently define risk to support a management decision, a higher assessment tier may be needed that may require either additional data or applying more refined analysis techniques to available data. Higher tiers provide more ecologically realistic assessments while making less conservative assumptions about exposure and effects.

Another option is to proceed with a management decision based on the risk assessment and develop a monitoring plan to evaluate the results of the decision, as noted in the example described earlier.

Communicating ecological risks to the public is usually the responsibility of risk managers. Although the final risk assessment document (including its risk characterization sections) can be made available to the public, the risk communication process is best served by tailoring information to a particular audience. It is important to clearly describe the ecological resources at risk, their value, and the costs of protecting (and failing to protect) the resources (US EPA 1995). The degree of confidence in the risk assessment and the rationale for risk management decisions and options for reducing risk are also important.

To date, the vast majority of risk assessments have addressed risks to individuals or groups of individuals (USEPA 1993; 1994a, b). However, there is considerable interest in assessing effects at higher levels of biological organization, e.g., at the population (Barnthouse et al. 1987), community (Brody et al. 1988), ecosystem, and landscape levels (Harwell et al. 1996). In addition, an important aspect of many ecological risk assessments is to assess the cascading of effects through the system (Lipton et al. 1993). For example, changes in the frequency of flooding may change the plant community, which in turn alters the wildlife populations that depend on the plants (Brody et al. 1988). Including such issues increases the ecological relevance of assessment as well as the chance that important ecological changes will not be overlooked, but also greatly increases the complexity and cost of performing the assessment.

Conclusions

The *Proposed Guidelines for Ecological Risk Assessment* (USEPA 1996) provides an analytical process that has been widely accepted as useful for performing risk assessment. Risk assessments may be designed to provide guidance to decision makers for a wide range of environmental alterations, including specific chemical stressors and physical habitat changes, or for differing combinations of multiple stressors. Ecological values, economics, and political factors all influence decision making, and designing risk assessments requires great flexibility. Data and information that may be useful in these more complex analyses include national risk-based data and site-specific risk information, in conjunction with regional evaluations of risk. As ecological risk assessment comes to be used more frequently to support landscape-scale management decisions, the diversity, breadth, and complexity of the risk assessments will increase significantly and may include evaluations that focus on understanding ecological processes influenced by a diversity of human actions and management options.

Acknowledgments

We gratefully acknowledge the efforts of our co-workers who contributed to the development of the *Proposed Guidelines for Ecological Risk Assessment*: Bill van der Schalie, Suzanne Macy Marcy, Pat Cirone, Don Rodier, Anne Sergeant, and Steve Wharton.

References

Barnthouse, L.W., G.W. Suter II, A.W. Rosen, and J.J. Beauchamp. 1987. Estimating responses of fish populations to toxic contaminants. *Journal of Environmental Toxicology and Chemistry* 6: 811-824.

Brody, J.S., W. Conner, L. Pearlstine, and W. Kitchens. 1989. Modeling bottomland forest and wildlife habitat changes in Louisiana's Atchafalaya Basin. In R.R. Sharitz and J.W. Gibbons (Eds.), *Freshwater Wetlands and Wildlife*. US Department of Energy Symposium Series (61), CONF-8603100, Office of Science and Technical Information, US Department of Energy, Oak Ridge, TN.

Harwell, M.A., J.F. Long, A. Bartuska, J. Gentile, C.C. Harwell, V. Myers, and J.C. Ogden. 1996. Ecosystem management to achieve ecological sustainability: the case of South Florida. *Environmental Management* 20: 497-521.

Holling, C.S. (Ed). 1978. *Adaptive Environmental Assessment and Management.* Chichester, UK: John Wiley and Sons.

Lipton J., H. Galbraith, J. Burger, and D. Wartenberg. 1993. A paradigm for ecological risk assessment. *Environmental Management* (17): 1-5

Moore, D.R.J., and G.R. Biddinger. 1995. The interaction between risk assessors and risk managers during the problem formulation phase. *Journal of Environmental Toxicology and Chemistry* (14): 2013-2014.

Suter, G.W. II. 1990. Endpoints for regional ecological risk assessments. *Environmental Management* (14): 19-23.

US Environmental Protection Agency (USEPA). 1992. *Framework for Ecological Risk Assessment.* EPA/630/R-92/001, Risk Assessment Forum, US Environmental Protection Agency, Washington, DC.

USEPA. 1993. *A Review of Ecological Assessment Case Studies from a Risk Assessment Perspective.* EPA/230/R-92/005, Risk Assessment Forum, US Environmental Protection Agency, Washington, DC.

USEPA. 1994a. *A Review of Ecological Assessment Case Studies from a Risk Assessment Perspective.* EPA/630/R-94/009, Risk Assessment Forum, US Environmental Protection Agency, Washington, DC.

USEPA. 1994b. *Managing Ecological Risks at EPA: Issues and Recommendations for Progress.* EPA/600/R-94/183, Office of Research and Development, US Environmental Protection Agency, Washington, DC.

USEPA. 1995. Memo to EPA managers from administrator Carol Browner, (March 1995). *EPA Risk Characterization Program*, US Environmental Protection Agency, Washington, DC.

USEPA 1996. *Proposed Guidelines for Ecological Risk Assessment.* EPA/630/R-95/002B, US Environmental Protection Agency, Washington, DC.

USEPA. 1997a. *Ecological Risk Assessment Guidance for Superfund: Process for Designing and Conducting Ecological Risk Assessments.* EPA 540-R-97-006, Office of Solid Waste and Emergency Response, US Environmental Protection Agency, Washington, DC.

USEPA. 1997b. *Priorities for Ecological Protection: An Initial List and Discussion Document for EPA.* EPA/600/S-97/002, Office of Research and Development, US Environmental Protection Agency, Washington, DC.

Ecological Risk and Policy Choice

Mark R. Powell[1]

Abstract

Ecological risk assessment presents policy problems that require resolution by political and social agreement. There is no scientifically correct answer as to what constitutes ecosystem health. Relying too heavily on either stakeholder consensus or science/scientists to resolve this dilemma runs the risk of abdicating tough policy choices to politically unaccountable decision makers. A robust procedure for analyzing and characterizing expert opinion regarding what constitutes an adverse ecological change is proposed as one possible means of achieving ecosystem protection decisions informed by science and judgment.

Introduction

Identifying ecological risk assessment endpoints (the environmental values to be protected) and defining what constitutes an adverse effect on poorly defined "ecosystem health" are "wicked" decision problems. Problems of the wicked sort are not necessarily complex or bounded by fuzzy or uncertain constraints, though they may be. Rather, they require resolution by political and social agreement. Absent such agreement, wicked problems remain – *de facto*, if not *de jure* – insoluble. (Legislative and judicial deadlines and decrees frequently require administrative policymakers to make decisions in the absence of consensus.) In environmental health risk assessment, analysts enjoy a broad consensus that human morbidity and

[1]Fellow, American Association for the Advancement of Science (AAAS), US Department of Agriculture (USDA), Office of Risk Assessment and Cost Benefit Analysis, Washington, DC, 20250-1000.

mortality are undesirable. Likewise, it is generally agreed that system failure is a bad thing in engineering risk assessment. Ecological risk analysis has no such compass by which to orient itself. Although there is some rhetorical value to the term – which should not necessarily be dismissed out of hand – there is no scientifically correct answer as to what constitutes ecosystem health.

There are at least two models for making choices about ecosystem protection policies: convening stakeholders to achieve consensus, and "objective" scientific decision making. Strictly following either model, however, invites abdication of tough policy choices by politically accountable decision makers. Ideally, we desire policymakers to make reasoned decisions informed by both science and value judgments. Before concluding by proposing a robust statistical procedure as one possible means of approaching this ideal, this discussion turns to a brief analysis of the "stakeholder-driven" and "science-driven" models.

The Stakeholder Model

A number of observers, including myself, have the impression that, by and large, the environmental governing philosophy across the Clinton administration has been to eschew policy analysis and to seek consensus by convening stakeholders around the bargaining table. Examples include "The Forest Conference" convened to resolve conflicting interests in the Pacific Northwest, the Habitat Conservation Plan initiative, designed to broker regional, multispecies management plans without resorting to the Endangered Species Act, the 1994 Executive Order on Environmental Justice requiring community participation in rulemaking and enforcement processes, and the participatory initiatives (the sector-specific Common Sense Initiative and the firm-specific XL) of the Environmental Protection Agency (EPA). Further support is provided by the decision to eliminate the independent, internal scientific and economic regulatory analysis functions previously conducted by EPA's policy and research offices. These initiatives may be interpreted as manifestations of an administrative strategy to substitute stakeholder consensus for science as the means of legitimizing environmental policy decisions. If the perception is indeed accurate that the current administration places a decreased emphasis on the use of "objective analysis", then this development may reflect some healthy skepticism about what "objective analysis" can contribute to value-laden environmental policy decisions. I would also argue that it represents a correction from the overselling of "science-driven" regulatory decision making by influential voices in the scientific community and by preceding administrations which sought to use the trappings of science to legitimize their policy choices.

Certainly, public input needs to inform ecological risk assessment problem formulation, and a strong consensus on ecological goals would seem to make it more likely that risk-management decisions will be supported during program implementation. The facilitation of conflict resolution, consensus building, and the sense of shared ownership that comes with participating in problem-solving are regarded by many observers as essential components in seeking solutions to environmental and natural resource policy problems. Public input, however, is no substitute for making tough policy choices. As inclusive as the administration's convening processes have appeared, some interests still feel left out or unplaced. One should be extremely skeptical about the potential to forge a meaningful consensus from a diverse public with conflicting or competing values. Additionally, we should be very cautious about abdicating regulatory decision-making authority to interested and affected parties. Priority-setting in particular can become extremely difficult as the number of parties who feel they have a legitimate claim to program resources increases. Unfortunately, as game theory suggests, win-win solutions are not possible in all cases. As Fisher and Ury (1981) observe:

> *However well you understand the interests of the other side, however ingeniously you invent ways of reconciling interests, however highly you value an ongoing relationship, you will almost always face the harsh reality of interests that conflict. No talk of "win-win" strategies can conceal that fact.*

As unsatisfied stakeholders still have access to a variety of veto measures, the current administration has been confronted with the limits of environmental policymaking by consensus.

The Scientific Model

Consider the case of identifying what constitutes an adverse ecological effect. Historically, the "ecological" effects of public concern have included conspicuous fish and bird kills. In some cases, however, these episodes may have no lasting impacts on animal population levels, the composition of ecological communities, or the structure and function of ecosystems. Despite public outrage or concerns about tangible welfare effects (consider the financial impacts of a noxious fish kill on eateries that depend on a pleasant lakeside environment), these events may be "ecologically irrelevant." But what does science have to say about determining the adversity of observed ecological changes?

Harwell et al. (1994) provide EPA unavoidably ambiguous scientific guidance on determining the significance of ecological changes. For an ecological change to be judged significant, according to Harwell and colleagues, it must exceed some estimate of natural variability, but statistical significance need not be demonstrated. Another way of phrasing this is simply to say that the 95-percent statistical confidence level is not divinely ordained. In its draft ecological risk-assessment guidelines, the agency offers a collective shrug, stating that determining ecological adversity "is not always an easy task and frequently depends on the best professional judgment of the risk assessor" (EPA 1996). In the regulatory context, the determination of an allowable rate of false negatives/positives is a policy judgment that should be informed by science. Risk managers may choose to follow scientific conventions for the sake of consistency or to avoid *ad hoc* decisionmaking, but there is no "scientifically correct" or purely "objective" answer. One questions, therefore, whether ecologists and risk analysts should have a monopoly on determining ecological adversity.

An analysis of the "Ecological Significance Framework" proposed by Harwell et al. (1994) (Figure 1) underscores that subjective judgments are inherent in making a determination about the significance or adversity of an environmental change. (Alternatives to Harwell's framework are indicated in Figure 1 by dashed lines.) It is not self-evident, for example, why a reversible effect of long duration, large scale, and small magnitude should necessarily be considered a significant change. An example of this sort would be a small, long-term increase in global surface temperatures, the significance of which is currently being hotly debated in the policy domain. (Personally, I would *advocate* that the specter of global climate change warrants significant precautionary investments in adaptation and mitigation, but note my choice of verbs.) Similarly, it is not readily apparent why the magnitude of the effect should not be taken into consideration for non-reversible, large-scale changes. If such a change were of relatively small magnitude, a decision maker might choose instead to acquire additional information before making a final determination.

Harwell and colleagues' framework reemphasizes that discerning whether a change exceeds some estimate of natural variability is critical to a determination of ecological adversity. They suggest that a default position is to assume that exceedance of natural variability is the norm rather than the exception; then, based on that assumption, one may proceed through the ecological significance decision tree and reassess variability only if it is deemed to be the critical component in the decision. An alternative default, however, would be to consider any determination of ecological significance to be provisional subject to a reassessment of variability. This discussion is not intended to criticize Harwell's proposed framework or its implicit judgments. Rather, the purpose is to underscore the degree to which risk assessors and risk managers must exercise judgment in determining what constitutes an adverse ecological effect. As suggested above, this decision may be scientifically elaborate, but it is ultimately a policy call.

A Robust Procedure for Evaluating and Characterizing Expert Opinion Regarding Ecological Significance

So, to review, both the stakeholder-driven and science-driven models are found lacking as the sole means of making public choices about ecosystem protection. I offer here one possible approach to combining science and judgment to inform environmental policy decisions.

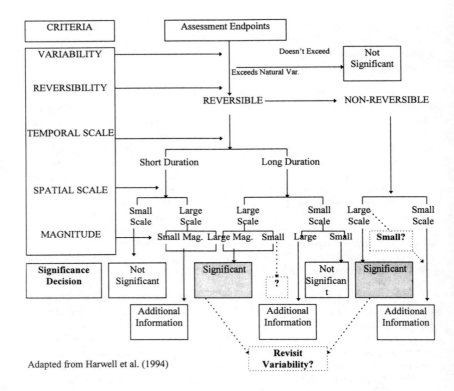

Adapted from Harwell et al. (1994)

Figure 1. Analysis of Harwell's Framework for Ecological Significance

The proposed procedure is a decision analysis model employing a robust confidence interval for location, based on a Descending M-estimator. It is intended for use in evaluating and characterizing expert opinion about what constitutes an adverse environmental effect. The procedure would be applicable only in cases where environmental responses to a stressor can be quantitatively characterized along a single, continuous dimension (e.g., percent change in a wildlife population level or soil loss per unit area from erosion). The procedure also would be limited to cases where there are a substantial number of potential "experts" to survey (e.g., 50 Natural Resource Conservation Service state conservationists).[2] Because determining ecological adversity combines scientific and value judgments, the situation is admittedly ambiguous. Thus, the expert survey respondents may be interpreted as

---

[2]To take an extreme case, if there were only two respondents, their responses could be arbitrarily close, and the M-estimator algorithm would fail to converge to a unique estimate of central location.

either a sample, probably non-representative, of the general population, or as a significant portion of the small population of qualified experts.

The model is intended to: 1) identify an "efficient" trapezoidal uncertainty distribution (i.e., as far from a uniform distribution as is justifiable) to summarize survey responses for communication to decision makers when an expert consensus exists; 2) identify when insufficient expert consensus may exist to justify combining the survey results into a single uncertainty distribution (i.e., when the Descending M-estimator algorithm fails to converge to a unique solution); and 3) safeguard against potential motivational biases on the part of the expert respondents to furnish extremely non-central survey responses.

It seems likely that the distribution of expert opinion regarding what constitutes an adverse environmental effect would violate assumptions of normality and symmetry. In general, the maximum likelihood estimate (MLE) can be made robust, or modified to reduce its sensitivity to outliers and yet still retain high efficiency at the ideal, assumed model. Huber (1964) modified the sample mean (the MLE of the population mean: under the normal distribution) to minimize asymptotic variance, where the maximum is taken over all symmetric distributions which are close to the normal distribution. This estimator is now called Huber's M-estimator (Ruppert 1985). The M-estimator can be expressed as a weighted average and is solved as an iterative minimization problem.

M-estimators that assign no weight to all very large deviations have been recommended, since these ignore gross outliers rather than merely bounding their influence. Such estimators are called descending M-estimators. If the normal distribution is contaminated by an asymmetric distribution, the problem of estimating central location becomes somewhat ambiguous. Huber (1981) shows that the sample median solves the problem of minimizing the maximum asymptotic bias. However, the minimax risk strategy can be unnecessarily conservative, and it can be worthwhile to increase the maximum risk slightly beyond its minimax value to gain better performance of the estimator (i.e., less variance) at long-tailed distributions. This can be accomplished by using a descending M-estimator (Hµber 1981). Tukey's confidence interval for the sample median (based on Wilcoxon's signed rank statistic) also presumes a symmetric distribution (Hollander and Wolfe 1973). Therefore, a descending M-estimator is well suited for characterizing the central location of distributions that are somewhat non-normal or asymmetric.

The robust confidence interval procedure developed by DuMond and Lenth (1987) is based on a descending M-estimator (the biweight), has high efficiency in the normal case, and maintains high validity over a broad range of distributions (uniform, symmetric heavy-tailed, and skewed heavy-tailed).[3] The interval is

---

[3]The procedure becomes somewhat unsafe when the population is both skewed and light-tailed.

expressed as:

$$M \pm t_{1-\alpha/2;\, n-1}[s/(\Sigma_n W_i)^{\frac{1}{2}}]$$

where $t_{1-\alpha/2;\, n-1}$ is the $(1-\alpha/2)$th quantile of the t distribution with n-1 df and s is the estimated standard error.

The biweight function (also called the bisquare) is given by:

$$w(t) = \{1 - (t/k)^2\}^2 \ \text{if } |t| < k$$
$$= 0 \qquad\qquad \text{if } |t| \geq k$$

where t is a standardized deviation from central location. Using a tuning constant (k) of 4.69 makes the efficiency of the biweight M-estimate relative to the sample mean (as a ratio of asymptotic variances) equal to 0.95 when the population is normal (DuMond and Lenth 1987). Table 1 compares some weights assigned by the biweight M-estimator (k = 4.69) with Huber's M-estimator (k = 1.5) and a trimmed mean (with w(t) = 0 for $|t| \geq 3$). Although the biweight function assigns zero weight to some extremely non-central (in this case, $t \geq 4.69$) responses to offset the risk of motivational bias, because the weights assigned by the biweight function decline smoothly and continuously, it can be regarded in some respects as a "fairer" rule for downweighting non-central observations (i.e., survey responses) than a discontinuous function (Huber's M-estimator or the trimmed mean).

Table 1. Comparison of Weights (w(t)) Assigned by the Biweight,
Huber's M-Estimator, and Trimmed Mean

t	Biweight	Huber's	Trimmed
0.00	1.00	1.00	1.00
1.00	0.91	1.00	1.00
1.50	0.81	1.00	1.00
2.0	0.67	0.75	1.00
3.00	0.35	0.50	0.00
5.00	0.00	0.30	0.00

The robust model for evaluating expert opinion would identify a trapezoidal uncertainty distribution to summarize survey responses regarding what constitutes an adverse environmental effect for communication to decision makers. In order to reduce the risk of information-overload and map down the problem for decision makers, the procedure is intended to characterize uncertainty about the "best guess" across experts. This is in the event that there is sufficient consensus among experts to justify combining the survey results into a single uncertainty distribution. However, it could be used alone or in combination with more conventional depictions of uncertainty within experts (individual subjective probability judgments expressed as a sigmoid-shaped, cumulative function) and among them (i.e., a series of such curves).

An attractive feature of the trapezoidal distribution (see Figure 2 below) is that it is more informative than a uniform distribution and (unlike the triangular distribution) does not require the specification of a singular modal value. However, specifying the endpoints (a,d) and turning points (b,c) of the distribution with precision often requires more information than may be available to the analyst (Seiler and Alvarez 1996). To identify the range of equally most-likely values in the trapezoidal uncertainty distribution, the model employs the confidence interval of location described above. The efficiency and robustness of the M-estimator procedure ensures that the trapezoidal distribution will be as informative (i.e., as far from a uniform distribution) as is appropriate while maintaining good validity over a broad range of underlying distributions of expert opinion. The endpoints of the distribution (a,d) would be the most extreme (i.e., non-central) non-zero weighted responses. The turning points of the distribution (b,c) would be the $\alpha/2$th and the $(1-\alpha/2)$th percentiles of the distribution.

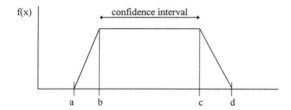

Figure 2. Probability Density Function for a Trapezoidal Distribution Describing the Distribution of Expert Opinion on the Percent Lung Function Decrement Regarded as Adverse

Valid objections can be raised about assigning variable "expert weights" to pool expert responses into a combined uncertainty distribution. Genest and Zidek (1986) demonstrate that imposed consensus functions require "dictatorial" aggregation methods ($w_i = 1$ for some i and 0 for others) and note their "impossibility" due to their unsatisfactory nature. However, the proposed model for evaluating expert opinion does not impose consensus. Instead, it provides a means of identifying whether expert consensus occurs, and if so, describing that consensus in a manner that is both informative (i.e., efficient) and robust (i.e., to deviations from the normal model and to motivational biases).

With a descending M-estimator, there will be multiple solutions to the minimization problem when there is insufficient central mass in the weighted probability distribution. Recall that computation of an M-estimate is an iterative minimization problem. If M's loss function is convex (successive iterations converge) with respect to the summation on the vertical axis of Figure 3, the minimizer may or may not be unique. If M's loss function is concave with respect to the summation, the minimizer is definitely non-unique. (The numbers in the graphs in Figure 3 refer to the iterative sequence.)

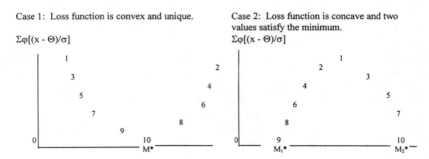

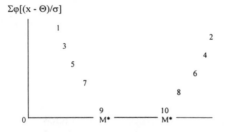

Figure 3. M-Estimate Uniqueness

While this ambiguous property of descending M-estimators may be a disadvantage in some contexts, it is a strength in the context of evaluating expert opinion because it means that the procedure does not impose a consensus where one may not exist. Conversely, when the procedure does yield a unique solution, there is increased confidence that an expert consensus can be identified and described in a manner that will not overload the decision maker with information. An explicit determination of a lack of consensus among experts can be an important finding by itself. If there is a lack of consensus among the experts regarding what constitutes an adverse environmental effect, one possible explanation is that experts are using quantitatively or qualitatively different biological models in forming their responses. Alternatively, it may signify polarized value-based opinions in the expert subpopulation. It should be noted that a determination of expert consensus could also reflect shared value-based opinions in the subpopulation.

In terms of addressing potential sources of bias among experts, existing uncertainty and decision analysis methods focus primarily on preventing or correcting for expert overconfidence (see Morgan and Henrion 1990). To counteract potential motivational biases, analysts may rely on procedural means to assure "balance" among experts. However, such efforts to avoid biasing the central tendency of expert opinion can easily bias the spread of the distribution. This proposed methodological approach could be used in conjunction with procedural means for addressing motivational bias in eliciting expert judgment to provide a robust and efficient description of the distribution of expert opinion regarding what constitutes an adverse ecological effect.

Note

This paper was presented while on leave of absence from the Resources for the Future (RFF), Center for Risk Management, Washington, DC, 20036. The views expressed are those of the author and do not reflect the position of AAAS, USDA, or RFF.

References

DuMond, C., and R. Lenth. 1987. A robust confidence interval for location. *Technometrics* 29(2): 211-219.

EPA (US Environmental Protection Agency). 1996. *Proposed Guidelines for Ecological Risk Assessment*. EPA/630/R-95/002B, August.

Fisher, R., and W. Ury. 1981. *Getting to Yes: Negotiating Agreement Without Giving In*. New York, NY: Penguin Books.

Genest, C., and J. Zidek. 1986. Combining probability distributions: a critique and an annotated bibliography. *Statistical Science* 1(1): 114-148.

Harwell, M., J. Gentile, B. Norton, and W. Cooper. 1994. Ecological significance. *Ecological Risk Assessment Issue Papers*. EPA/630/R-94/009 (2): 1-49, November, US Environmental Protection Agency Risk Assessment Forum.

Hollander, M., and D. Wolfe. 1973. *Nonparametric Statistical Methods*. New York, NY: John Wiley and Sons.

Huber, P.J. 1964. *Annals of Mathematical Statistics*. 35: 73-101.

Huber, P.J. 1981. *Robust Statistics*. New York, NY: John Wiley and Sons.

Morgan, G., and M. Henrion. 1990. *Uncertainty: A Guide to Dealing with Uncertainty in Quantitative Risk and Policy Analysis*. New York, NY: Cambridge University Press.

Ruppert, D. 1985. M-estimators. In S. Kortz and N. Johnson (Eds.), *Encyclopedia of Statistical Sciences* 5. New York, NY: John Wiley and Sons.

Seiler, F., and J. Alvarez. 1996. On the selection of distributions for stochastic variables. *Risk Analysis* 16(1): 5-18.

Extreme Value Analysis of Wastewater Treatment Plant Flows

Jonathan W. Bulkley[1]

Abstract

The implementation of the Clean Water Act has resulted in the provision of enhanced wastewater treatment facilities throughout this country. The initial effort focused upon providing a minimum of secondary treatment (or higher-order treatment if needed) to meet surface water quality standards under dry weather flow conditions. Since 1987, increased emphasis has been placed by both federal and state regulatory agencies upon meeting effluent limits from these treatment facilities under wet weather flow conditions. In communities served by combined sanitary and stormwater sewers, the treatment plant may experience very significant variations in flows to be treated at the facility. This paper will present both a Gumbel Type I extreme value distribution and a Gumbel Type II extreme value distribution to examine the probability of high flow exceedance at a treatment facility. This type of analysis provides a preliminary framework for consideration of the trade-off between fully capturing extreme value flows vs. occasionally by-passing extreme value flows with less than full treatment.

Introduction

The implementation of the Clean Water Act since 1972 has resulted in the enhanced performance of publicly owned wastewater treatment plants (POTWs) throughout this country. Initial efforts focused upon undertaking the necessary construction projects and improving operations to ensure that the effluent discharges

[1]Professor of Natural Resources, Professor of Civil and Environmental Engineering, University of Michigan, School of Natural Resources and Environment, Ann Arbor, MI 48109-1115

from those facilities were in full compliance with all federal and state water quality standards in the receiving water during dry weather flow conditions. The expenditure of more than $58 billion for needed construction and improvements to POTWs has demonstrated that these facilities can be designed, built, and operated as specified in their discharge permits to meet their effluent limits during dry weather flow conditions.

Biannual water-quality surveys of the surface waters of the United States demonstrate that roughly 2/3 of the surface waters are meeting designated standards. Where there is non-compliance, the major contributing factors are overflows and discharges from combined sewer systems, storm water systems, and non-point source pollution reaching the receiving water. Wet weather flow conditions are one major element leading to the impairment of surface water quality in the United States.

Elements of the Problem

The response of a sewage collection and treatment system to wet weather flow events is a function of many variables, including the spatial characteristics of the collection systems, whether the collection systems include combined or separate sanitary sewage and storm water overflows, the age of the system, the maintenance and operational reliability of the system, the land use(s) of the service area, the infiltration and inflows that impact the collection system, and the management practices that are in place, especially for street cleaning and other pollution-prevention activities in the service area.

The POTW itself is the processor of sanitary sewage and the storm water that has mixed with it, either by design as in the case of a combined sewer system, or by failure through excessive inflow and infiltration into the collection network of interception sewers. This network brings the wastewater to the plant for treatment prior to its discharge into the receiving waters. Under high flow conditions following wet weather events or snow melt, the capacity of the POTW to provide the desired full treatment to all flows is limited. For example, a full secondary treatment facility may be able to provide complete secondary treatment for flows only up to a fixed design flow; flows in excess of this capacity may receive primary treatment and disinfection prior to discharge into receiving waters. Accordingly, it is important to recognize that a complex set of factors combine to determine the volume and strength of the wastewater reaching a POTW following wet weather events or snow melt.

Although all of these issues are important, this paper will focus specifically on utilizing extreme-event analysis to examine daily flows observed at a POTW. Using this approach, it becomes feasible to examine the probability of flow exceedance at a particular plant on a daily basis for extended periods of time. Extreme-event analysis should contribute to the discussion of trade-offs between increasing capital investment(s) to capture and provide full treatment for increasingly rare-event situations vs. capturing the first major flow and then providing for partial treatment of high-flow/low-frequency discharge events into receiving waters. Applying the techniques of extreme-event analysis will not fully resolve the issue; however, the application of these techniques should contribute to a more informed discussion among all parties, including the regulators, the operators of POTWs, environmental groups, and the public.

The Basic Problem

A regulatory agency seeks to have each POTW capable of providing full treatment to flows in excess of normal dry weather flow conditions. Space limitations at existing POTW locations often constrain the physical facilities which can be built on the site. For example, if a POTW normally treats a flow of 70 mgd, at times it may be required to provide full treatment for extreme wet weather flows up to 160 mgd, a flow greater than twice its dry weather flow. However, there may be physical limits in terms of space at the existing treatment plant. Or, the limits may be economic, in terms of what constitutes a cost-effective allocation of limited financial resources to build and maintain excessive additional capacity to capture and treat excessively high flows which occur at very low frequency. Extreme-event analysis provides a method for considering the probabilities of occurrence of these extremely high-flow events associated with wet weather conditions and snow melt.

Extreme-Event Analysis: Water Reclamation Plant

Identify and select the maximum 24-hour flow which is shown on each monthly operations report (MOR) form prepared by the operators of the treatment plant. (111 MOR forms were examined from October 1986 to August 1997.)

Type I extreme value distribution: sample calculation

- Use the Gumbel Type I extreme value distribution (Benjamin and Cornell 1970).

  Cumulative distribution function: $F_y(Y) = \exp\left[-e^{-\alpha(y-\mu)}\right]$

  y = random variable

In this case, the random variable is the flow in million gallons per day (mgd) metered at the plant over a 24-hour period. By choosing the maximum 24-hour flow in each month of record, one obtains a new distribution of extreme values:

$$M_y = \mu + \frac{0.577}{\alpha}$$

where M_y = the mean of the distribution of observed maximum flow, and μ = the mode of the distribution.

where M_y = the mean of the distribution of observed maximum flow, and μ = the mode of the distribution.

$$\sigma_y = \frac{1.282}{\alpha}$$

where σ_y = the standard deviation of the distribution of observed maximum flows. Note that the parameters μ and α are obtained from the mean and standard deviations of the new distribution of extreme flow values.

$$M_y = 129.42 \text{ mgd}$$
$$\sigma_y = 39.57 \text{ mgd}$$

Therefore,

$$\alpha = \frac{1.282}{39.57} = .0324$$
$$\mu = M_y - \frac{.577}{\alpha} = 129.42 - \frac{.577}{.0324}$$
$$\mu = 129.42 - 17.81 = 111.61 \text{ mgd}$$

Now one may proceed to calculate the probabilities of exceedance of any desired flow, and conversely, calculate the flow associated with any desired probability of exceedance.

Figure 1 shows the cumulative probability function for this Type I Extreme Value Analysis. Table 1 shows the results from calculating the $F_y(y)$ for a number of extreme value flows.

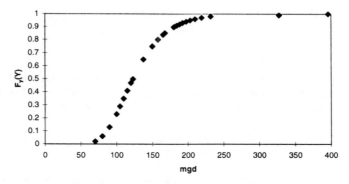

Figure 1. CDF Gumbel Type I Extreme Value Distribution

Table 1. Type I Extreme Value Distribution:
Maximum 24-Hour Flow Treated at the Reclamation Plant

		y	$F_y(Y)$	$1 - F_y(Y)$
		mgd	$P[y \leq Y]$	$P[Y \geq y]$
		396	.9999	.0001
		232	.98	.02
		203	.95	.05
City	→	180*	.897	.103
U.S.	→	165	.838	.162
		181	.90	.10
		168	.85	.15
		158	.80	.20
		150	.75	.25
		137	.65	.35
Median		123	.50	.50
Mode		112	.37	.63

*Solve for $F_y(Y)$ when $y = 165$ mgd and when $y = 180$ mgd.

$$F_y(Y) = \exp\left[-e^{-\alpha(y-\mu)}\right] \quad \begin{array}{l} \alpha = .0324 \\ \mu = 111.61 \text{ mgd} \end{array}$$

Type II Extreme Value Distribution: sample calculation

- Use the Gumbel Type II Extreme Value Distribution. Limited at left by zero: cannot be negative. Unlimited at right → tail of interest (Benjamin and Cornell 1970).

Cumulation Distribution Function: $F_y(Y) = e^{-(\mu/y)^k}$

y = random variable

Again, the random variable is the flow in million gallons per day (mgd) metered at the plant over a 24-hour period. This time, the maximum 24-hour flow in each month of record yields a new distribution of extreme values:

$$M_y = \mu\Gamma\left(1 - \frac{1}{K}\right)$$

where M_y = the mean of the distribution of observed maximum flow, μ = the mode of the distribution, k is a parameter obtained from the coefficient of variation (CV) $\dfrac{\sigma_y}{M_y}$, and a plot of the Type II Extreme Value Distribution.

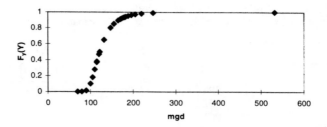

Figure 2. Gumbel Type II Extreme Value Distribution

Table 2. Type II Extreme Value Distribution:
Maximum 24-Hour Flow Treated at the Reclamation Plant

		y	$F_y(Y)$	$1 - F_y(Y)$
		mgd	$P[y \le Y]$	$P[Y \ge y]$
		532	.9999	.0001
		220	.98	.02
		188	.95	.05
City	$\rightarrow$	180*	.936	.064
US EPA	$\rightarrow$	165	.894	.106
		171	.90	.10
		155	.85	.15
		147	.80	.20
		141	.75	.25
		132	.65	.35
Median		122	.50	.50
Mode		115	.37	.63

*Solve for $F_y(Y)$ when $y = 165$ mgd and when $y = 180$ mgd.

$$F_y(Y) = e^{-\alpha(\mu/y)^\kappa} \qquad \begin{array}{l} K = 6.0 \\ \mu = 114.65 \text{ mgd} \end{array}$$

In this specific case, $M_y = 129.42$ mgd and $\sigma_y = 39.57$ mgd; accordingly, $CV = \dfrac{39.57}{129.42} = .3057 = .31$; and $K = 6$ (Benjamin and Cornell 1970). The value of the gamma distribution is obtained from appropriate tables (Burington and May 1950).

Accordingly, the mode, μ, is calculated as follows: $\mu = \dfrac{129.42}{1.13} = 114$ mgd .

Next, return to $F_y(Y) = e^{-(\mu/y)^K}$ and proceed to calculate the probabilities of exceedance of any desired flow. Conversely, calculate the flow associated with any desired probability of exceedance.

Figure 2 shows the cumulative probability function for this Type II Extreme Value Analysis. Table 2 shows the results from calculating the $F_y(y)$ for a number of extreme value flows.

Observations

Gumbel Type I and Type II Extreme Value Distributions may be used to sharpen understanding and provide insight into the probabilities associated with flow exceedance during wet weather events at POTWs. In the specific case example:

1) The regulatory agency desires that the POTW treat all flows ≤ 165 mgd to full secondary treatment.
2) The operator of the POTW stated that it would **comply** and believes it can provide full secondary treatment for all flows ≤ 180 mgd. Furthermore, the owner/operator would agree to stipulated penalties if the facility failed to meet its effluent limits for any flow ≤ 180 mgd.
3) The regulatory agency desires that the POTW build additional treatment capacity to ensure that it will be able to perform under wet weather/snow-melt conditions.
4) i.   Using the Type I Extreme Value Distribution for this POTW:

	Flow	$F(Y)$	$1-F(Y)$
Regulatory	165 mgd	.838	.162
POTW	180 mgd	.897	.103

   ii. Using the Type II Extreme Value Distribution for this POTW, probability of flow $\leq$ specified flow, $Y = F(y)$:

	Flow	$F(Y)$	$1-F(Y)$
Regulatory	165 mgd	.894	.106
POTW	180 mgd	.936	.064

5) Note the differences especially in the probability of exceedance $[1-F(y)]$.
6) Check the probability of exceedance against the 111 months of daily data by counting the observed numbers of flows that exceed 165 mgd and 180 mgd.

	Observed data ( )	≥ 165 mgd	≥ 180 mgd
	( )	(22)	(12)
Predicted - Type I		18	12
Predicted - Type II		12	7

7) Type I extreme value distribution appears to provide enhanced insights for this specific example.

8) Note: months of year of flow exceedance > 180 mgd.

October	2	9.25 years
November	3	
January	1	
⎡ April	1 ⎤	
│ May	1 │	
⎣ June	2 ⎦	
February	1	
December	1	
	12	

Namely: the observed flow at the treatment plant exceeded 180 mgd a total of 12 times in 9.25 years. During warm weather, there were four (4) flows over a 24-hour period that exceeded 180 mgd.

Conclusions

Gumbel Type I and Gumbel Type II extreme value distributions have been applied to wet-weather flow conditions at a wastewater treatment plant. Results from this example, which used 9¼ years of daily data, indicates that the Gumbel Type I extreme value distribution produced results closer to the results observed from the actual flow data at the facility.

In this specific case, the operator of the POTW asserted that the facility could treat flows up to 180 mgd without violating its NPOES permit. The Gumbel Type I extreme-event analysis indicates that in a 30-day (Monthly Operating Report) period, the probability of having a flow that exceeds 180 mgd is .106. Over 111 months of record, this exceedance probability indicates that one could expect 12 flows over 180 mgd during a 24-hour period. The actual data contained 12 days of flows in excess of 180 mgd. Two-thirds of these flows were observed in either fall or winter months; one-third (4 days) were observed in spring and early summer.

Using this method, the issue for the regulatory agency and the operator of the POTW is more fully illuminated, at least in terms of complete capture of flows over 180 mgd. Should additional funds be expended so that flows in excess of 180 mgd can be captured for full treatment, or should the POTW be allowed to provide partial treatment for such flows? It becomes increasingly difficult to justify expenditures for major capital works whose uses may be limited because of the low frequency of events which require them.

References

Benjamin, J.R., and C.A. Cornell. 1970. *Probability, Statistics, and Decision for Civil Engineers*. New York, NY: McGraw-Hill.

Burington, R.S., and D.C. May. 1950. *Probability and Statistics*. Sandusky, OH: McGraw-Hill.

Adaptive Risk Analysis for Resource Conservation Programs

Ronald Meekhof[1], Jennifer Kuzma[2], David Mauriello[3],
Tim Osborn[4], Mark Powell[5], Cliff Rice[6], and Steven Shafer[7]

Abstract

The following paper presents a guiding strategy for the implementation and analysis of the United States Department of Agriculture's conservation programs. This strategy involves a process which we term adaptive risk analysis. The adaptivity of the process stems from important feedback and monitoring mechanisms during all stages of program development, which can then be used to incorporate changes into a particular program as it evolves. Adaptive risk analysis occurs in several stages: 1) program identification, 2) environmental risk assessment, 3) program analysis, 4) cost-benefit analysis, 5) implementation, and 6) monitoring and evaluation. National, state, and local-level issues and activities are described for the various stages, and communication among program managers at all levels is emphasized. Identification of program objectives, environmental indicators, mitigation measures, and costs-benefits are essential components of the process.

[1]Deputy Director, Office of Risk Assessment and Cost-Benefit Analysis, US Dept. of Agriculture, Washington DC 20250
[2]AAAS Fellow, Office of Risk Assessment and Cost-Benefit Analysis, US Dept. of Agriculture, Washington, DC 20250
[3]Senior Ecologist, US Environmental Protection Agency, 401 M Street SW, Washington, DC 20460
[4]Deputy Director, Resource Economics Division, Economic Research Service, 1800 M Street NW, Washington, DC 20036
[5]Resources for the Future, 1616 P Street NW, Washington, DC 20036
[6]Environmental Chemistry, Agricultural Research Service, Beltsville, MD 20705
[7]Animal and Plant Health Inspection Service, Raleigh, NC 27695

Introduction

Adaptive risk analysis is a tool for structuring information about environmental hazards and their consequences, identifying sound programs to prevent or reduce those hazards, and for determining whether resource conservation program objectives are being achieved in a cost-effective manner. Adaptive risk analysis provides a basis for ongoing improvement in program management by providing feedback on performance in reducing environmental hazards. It provides a framework for the development and evaluation processes by identifying the technical and management linkages between objectives, program alternatives, assessment endpoints, monitoring, and program performance.

The Conservation Reserve Program (CRP), Environmental Quality Incentives Program (EQIP), and other US Department of Agriculture resource conservation programs promote the sustainability of natural resources, including the maintenance of ecological functions. Managing programs for this objective implies that decisions must consider the variability and the uncertainties associated with agricultural activities and resource conservation measures. By improving our understanding of risks and uncertainties, program managers can better understand the ranges of outcomes that can be expected and whether these ranges promote sustainable use (Cleaves 1995). Adaptive risk analysis relies on the use of reasonably available information, sound analytical and management methods, and monitoring to better understand the relationships between agricultural risks, risk-mitigation activities, and the ranges or likelihoods associated with program outcomes. Without such information, regulatory controls or prohibitions may be employed to reduce risks to unnecessarily low levels at unnecessarily high costs or with little incremental effect when a more flexible, voluntary approach may have been sufficient.

This paper presents a broad outline for the adaptive risk analysis approach to regulatory impact analysis of major rules issued by the USDA concerning environmental hazards. This preliminary guidance is being developed to show how the results of recently conducted risk assessments for the CRP and EQIP can be used to evaluate the environmental benefits of these programs, which are now being implemented.

The Adaptive Risk Analysis Process

The principal components of the adaptive risk analysis process are ecological risk assessment, risk management (which includes program analysis, cost-benefit analysis, and implementation), and monitoring and evaluation. An ecological risk assessment initiates the process by providing information on the likelihood of undesirable events or activities and the ecological consequences. The assessment should identify significant risks, important ecological relationships, and areas of uncertainty. It should also identify ecologically significant assessment endpoints (e.g., ambient water-quality criteria) which guide risk management as well as monitoring and evaluation. Risk management is the process by which suitable measures or program alternatives for reducing or preventing risks are identified and implemented. Program alternatives are evaluated on the basis of their effectiveness in reducing significant ecological hazards, the level of participation, and the degree of risk mitigation obtained. The magnitude of the expected benefits are compared to the economic, social, and other costs in the selection process. Areas of uncertainty and their impacts on the analysis and findings should be identified. The analysis should examine the types of resources affected and the overall improvements in environmental benefits, and the costs of achieving those benefits. Monitoring provides information on progress in achieving program objectives and feedback, for use in modifying program implementation.

The value of using the adaptive risk analysis method will be determined by the extent to which the best, most reasonably available scientific, economic, and technical information is utilized, and the extent to which the method is integrated with the management process of the organization. Agencies which manage programs from a central location are likely to use the adaptive risk analysis method in a very different way from agencies which rely on decentralized management. Decisions at the national level are primarily concerned with identifying program objectives, setting priorities among objectives, and targeting priority areas of national or regional importance. Analysis is primarily concerned with comparing the cost effectiveness of alternative programs.

On the other hand, decentralized or local decisions are targeted, and include the identification of appropriate risk management practices under specific environmental and crop and livestock production conditions. Analysis in support of local program decisions is necessarily more fine-grained. It can be based on the results of studies or models ranging in scope from the field to small watersheds. Monitoring and evaluation for local decision making may include on-farm and local off-farm observation, measurements, and analysis.

USDA resource conservation programs involve aspects of both centralized and decentralized management approaches. The following sections describe the components of the adaptive risk analysis process as it may be applied to these programs, and the types of issues likely to arise at the national, state, and local management levels.

Program Identification

The program identification component identifies the resource concerns and the management goals. A prospective risk assessment or conceptual model can provide a preliminary perspective on program objectives, types of mitigation activities, ecological relationships, and assessment endpoints. This component is also concerned with identifying broad management options and program constraints. Information needs, resources, and means for incorporating stakeholder participation should be identified as well (Commission on Risk Assessment and Risk Management 1997).

NATIONAL:

- What are the major resource concerns? What are the major programmatic approaches that can be used to address these concerns?

- Are there conflicting goals and objectives? How will such conflicts be resolved?

- If multiple objectives are to be achieved by the program, have priorities been identified?

- What are the critical administrative and programmatic concerns regarding program implementation?

- Are there other federal or state programs whose activities should be considered in program identification?

- What are the types of uncertainty of greatest concern to risk managers?

STATE:

- What are the major resource problems and conservation objectives in the state?

- Have stakeholders had the opportunity to express their views about the program and how to best achieve the objectives?

- Are there other environmental programs operating in the state where program performance can be enhanced through coordination?

LOCAL:

- Have local surveys been conducted which identify and document the resource problems, opportunities, and concerns of the program clientele? Has this information been communicated to the state and national levels?

- Have local stakeholder groups provided comments on the program? Are they reflected in program identification?

- Is there a long-term conservation plan for the local area and region which identifies conservation objectives and priorities?

Environmental Risk Assessment

The purpose of an agricultural ecological risk assessment is to identify the effects of hazards initiated by agricultural activities which pose the greatest threat to environmental conditions, including ecological and human health. The consequences of the hazards for ecological relationships and the corresponding assessment endpoints need to be identified using the best, most reasonably available scientific and technical information (USEPA 1997). The risk assessment should also identify areas of uncertainty. It should be structured in a way that provides the types of information needed to evaluate program alternatives. An important task of the risk assessment is to identify assessment endpoints that are ecologically significant relative to program objectives, are responsive to program mitigation activities, and can thereby measure program performance. The program identification component can help identify the questions to be addressed and the types of information needed.

Risk assessments have been conducted for the CRP (USDA 1997a) and EQIP (USDA 1997b) which identify the cause-and-effect pathways of environmental risks such as soil disturbances, irrigation water application, pesticide application, nutrient application, pasture and rangeland grazing, confined livestock production, and others. Assessment endpoints associated with risk-initiating activities have been identified. These general assessment endpoints include: structure of off-site resources and habitats, livestock or plant yields, wetland functions, viability of aquatic communities, good air quality, potable water supplies, terrestrial and avian wildlife communities, and threatened and endangered species. The cumulative effects of the hazards were also assessed qualitatively to identify regional areas where risks from agricultural activities may pose greater harm.

NATIONAL:

- Have assessment endpoints been identified which reflect the ecological consequences of the hazards addressed by the program? Will changes in the assessment endpoints reflect program-mitigation activities?

- Can the indicators be valued in terms of environmental and economic benefits?

- What types of changes in assessment endpoints could be expected within the range of natural variability?

- Has a plan for monitoring and evaluation been developed which reflects risk assessment results?

STATE:

- Are the results of the risk assessment consistent with an analysis of state resource concerns to be addressed by the program?

- Have assessment endpoints been selected which measure the effectiveness of the program in alleviating resource concerns in the state?

- Have areas in the state been identified which are significant because of cumulative risks or other factors?

LOCAL:

- Has sufficient information been gathered to analyze and understand the natural resource conditions in the area?

- Have the resource stressors been thoroughly identified and documented with regard to their impacts on resource concerns?

- Have ecological baseline conditions been established? Are there areas in the locality where the extent of the resource problems exceed the resources available through the existing or proposed program? What types of effects are observed?

- Are the results of the analyses reflected in national program objectives?

Program Analysis

The purpose of this component is to identify alternative policy approaches and program alternatives for mitigating hazards and achieving program objectives. An analysis of alternative policies examines broad approaches (land retirement, cost share) for achieving resource conservation objectives. An analysis of program alternatives may be limited by statutory requirements or policy. Such an analysis identifies options, evaluates their effectiveness in reducing harm to environmental resources, assesses their feasibility, and examines the unintended effects. Given the statutory requirements, the complexity of USDA resource conservation programs, decentralized decision making, and the need for flexibility, the identification of the alternatives may require an iterative process. Broad policy goals, program alternatives, and implementation issues may be modified based on the results of the analysis. Program evaluation may also reflect stakeholder concerns and the need for federal, state, and local program coordination.

Analysis of conservation programs, prior to implementation, is carried out at the national level, and regulatory and policy guidance is then provided to the state and local levels. Analysis may also be encouraged at the state level when there are significantly different resource concerns among the states.

Identification of program mitigation measures

A range of program alternatives should be identified on the basis of the risk assessment and program identification results. The alternatives generally differ in regard to the level of intervention, targeting, level of incentives, and other characteristics. The alternatives may also differ in their emphasis on program objectives. The identification of potential mitigation measures should reflect whether the program is mandatory or voluntary, cost-share, technical assistance, or otherwise.

Ecosystem characterization

The significant relevant ecological characteristics of the farm and off-farm ecosystems are identified. The hazards, ecological relationships, ecological effects, and assessment endpoints should be described in sufficient detail to identify measurable effects of program alternatives. Ecosystem models, ecosystem parameters, or other quantitative and qualitative information may be used to characterize the ecosystems. In many cases, the risk assessment will provide much of this information. However, additional information and modeling may be necessary to evaluate the impacts of program alternatives on ecosystem parameters.

Determination of environmental benefits

Using the ecosystem characterization information and models, estimate the expected reduction in risk for the program alternatives. The reduction in risk, or environmental benefit, of the alternatives is based on changes from baseline conditions. Changes in assessment endpoints indicate the types of environmental benefits. The analysis should focus on those assessment endpoints that are ecologically significant. It should indicate how changes in assessment endpoints are measured and what assumptions, if any, were used to derive environmental benefits. Other risk-reduction consequences can be discussed, including human health benefits and the development of new technologies.

Resource improvement characterization

Discuss the broader environmental, economic, and social implications of the program. Identify groups or regions that are likely to be most affected by the program alternatives. Summarize the types of resources affected by the program and the expected risk-reduction benefits based on results of the analysis phase. Discuss the sources and level of uncertainty associated with results of the program analysis. Incorporate uncertainty analysis into analytical models and parameters. Describe the range of expected outcomes for the program alternatives.

NATIONAL:

- Does the analysis provide a clear understanding of how the program alternatives will alleviate resource concerns?

- Were the stakeholders involved in identifying alternatives?

- Does the analysis adequately identify areas where cumulative risks may be a major concern?

- Are sources and implications of uncertainty identified and incorporated into the analysis?

STATE:

- Does the national program analysis need to be modified to reflect ecological and resource conditions in the state?

- What types of information and analytical resources are available to target regions in the state where resource concerns are greatest?

- Have stakeholders presented their views on program alternatives and implementation issues?

LOCAL:

- Do the alternatives allow sufficient flexibility in achieving program objectives?

- Have alternative mitigation measures been identified which meet client and program objectives?

- Have the significant hazards, ecological relationships, ecological effects, and assessment endpoints been adequately described and documented?

Cost-Benefit Analysis

Cost-benefit analysis (CBA) is a technique to improve the quality of program decisions by evaluating in quantitative and qualitative terms the economic implications of program alternatives (OMB 1996). A cost-benefit analysis enables risk managers to determine whether the expected costs of the program are justified by the benefits, and whether the program maximizes net benefits to society, is most cost-effective, adheres to performance objectives to the extent feasible, or meets some other regulatory standard. CBA can be used at the local level for identifying cost-effective alternatives for program clientele. The advantage of the technique is that highly dissimilar risk-reduction benefits and costs can be aggregated into a measure of net benefits. While there may be significant difficulties in aggregation, at a minimum CBA can provide a ranking of policies on the basis of social benefits.

The most important practices for conducting a CBA are the use of clear and consistent baseline assumptions, the evaluation of a reasonable set of program alternatives, appropriate methods for discounting future costs and benefits and accounting for the cost of bearing risk, the use of probabilistic analyses to assess the strength of the conclusions, the identification of quantifiable and nonquantifiable effects, and the identification of distributional effects (Kopp et al. 1996).

Program benefits

The information on environmental benefits and resource improvements characterization is used to evaluate the expected reduction in risk, or environmental benefits, for the program alternatives. These results are compared to the conditions expected in the absence of the program, or the baseline. The changes in assessment endpoints are translated into economic effects through the use of economic models and assumptions. The improvements in assessment endpoints are the best means for evaluating the program alternatives and need to be evaluated from an economic perspective. The analysis of program benefits may reflect other   information, including the extent of the hazard in the ecosystem or site being treated, the extent to which the program is utilized, and the quantitative relationship between the practices and the degree of mitigation obtained.

Program costs

The economic costs of a regulation are generally measured in terms of the opportunity costs of the resources used or the benefits foregone as a result of regulatory action. Private sector compliance costs and program administrative costs are opportunity-foregone costs which are most easily measured. Other costs include the value of benefits of a product or use of a resource, lost productivity, and employment losses. A slowing in the rate of technological change and innovation due to stringent standards is another type of cost. The effects of a regulation on technological change and employment are often difficult to measure. A regulation may also result in certain costs being avoided. These are measured as benefits. Sunk costs and transfer payments are not economic costs. Sunk costs are those which have already occurred; transfer payments are a reallocation of income from one group in the economy to another.

Preferred program alternative identification

Based on the analyses, the risk assessors, economists, and other analysts should identify the program alternative which is most effective, provides the greatest net social benefit, or best achieves some other stated criterion. The choice also reflects the relative effectiveness of the program alternatives to achieve stated conservation goals, the strengths and weakness of the information available, and uncertainty. The identification should be justified with relevant information on the types of results that can reasonably be expected to occur and their likelihood. It may also identify groups or regions of the country that may be particularly affected.

NATIONAL:

- Does the baseline reflect likely changes in the economy, the agricultural sector, technology, other resource conservation policies, and other factors affecting costs and benefits?

- Are the methods and assumptions used in placing economic values on the risk-reduction consequences of program options documented? Are qualitative effects discussed?

- Are the risks and uncertainties associated with the alternatives reflected in the analysis? Where uncertainty is significant, are there cost-effective means of obtaining additional information to reduce uncertainty?

- Are the improvements in assessment endpoints evaluated in terms that are meaningful for evaluating the performance of the program?

STATE:

- Does the preferred program alternative adequately reflect regional conservation planning initiatives?

- Are there special resource concerns in the region that are not adequately reflected in the preferred option?

- Does a proposed alternative have a disproportionate effect on a group or region in the state?

LOCAL:

- Does the analysis adequately reflect the expected risk-reduction benefits and costs of the program alternatives? What might cause significant differences?

- Does the preferred option enable local personnel to meet quality criteria and clientele objectives?

- Are the social and economic values of the local community adequately reflected in the analysis?

Implementation

The implementation component is comprised of deciding which program alternative is most appropriate for achieving the stated objectives, and then taking actions to put the selected program into effect. For national resource conservation programs which address multiple assessment endpoints, achieve multiple objectives, and operate in multiple ecosystems, the implementation phase can require extensive negotiation and compromise among risk managers, risk assessors, stakeholders, program managers, field staff, and others. The analysis conducted in previous stages helps identify and narrow the range of reasonable alternatives. The information resulting from the risk assessment and program analysis stages should help the risk manager identify which alternative is the most suitable for achieving program objectives, and the associated level of uncertainty. Implementation of flexible, voluntary programs, such as EQIP and CRP, requires a high level of coordination at the national, state, and local levels. Analyses of the types of problems which may arise at various steps in the implementation process and of the resource requirements may be required if the program is new or is being significantly modified.

The risk manager may also consider additional types of information resulting from stakeholder input, and political and legal considerations. Involving stakeholders in the analysis and decision-making process may provide guidance, improve implementation, and promote trust in agency analysis and operations.

Implementation should reflect local natural systems and ecological processes that sustain the resources treated under the program. Resource plans need to account for the characteristics and potential of the resources potentially enrolled in the program, and the landowner's physical and economic capacity to implement recommended program practices. Broader social and economic issues may enter the implementation process. Effective implementation may require consideration of other site-specific characteristics, such as the location of the enrolled parcel on the farm or within the watershed, or its proximity to sensitive habitat or land and water resources (USDA 1996).

NATIONAL:

- Were the program alternatives thoroughly discussed by risk assessors, risk managers, and others prior to selection?

- Were technical implementation issues at the state and local levels evaluated? Do they have a significant impact on the expected performance of the program alternatives?

- Does the level of uncertainty about program performance warrant further analysis?

STATE:

- Will the selected program enable states to achieve the resource conservation objectives identified in the state-wide conservation plans?

LOCAL:

- Do local offices have sufficient personnel and technical resources to carry out the program?

- Does the clientele have sufficient information to implement and maintain planned program-mitigation activities?

Monitoring and Evaluation

The successful implementation of conservation programs over time requires establishing institutional procedures for monitoring and evaluation. Such feedback must be integrated into the risk analysis framework to provide the necessary information for determining the effectiveness of program implementation. Monitoring provides the means to evaluate alternative management options and to determine the level of success in implementing resource conservation program objectives.

Monitoring and evaluation efforts should be tightly coupled with program objectives at each level at which the program is implemented – national, regional, and local. This implies selecting monitoring and evaluation endpoints that are most effective in measuring the risk-reduction benefits. An additional requirement is that the measurement endpoints identified must be scalable from local to national levels in order to provide for the aggregation of information collected. Establishing a baseline set of conditions and ranges to reflect natural variability is also critical. Thus, the establishment of an experimental design for monitoring and evaluation purposes, involving the selection of indicators and thresholds, should occur concurrently with the conceptual phase of program identification, at the same time as management objectives, assessment endpoints, and implementation options are selected.

For convenience, monitoring activities may be subdivided into three major categories: implementation monitoring, effectiveness monitoring, and validation monitoring. The relationship between the phases of the adaptive risk analysis program and the monitoring efforts is shown in Figure 1 and is described below:

Implementation monitoring is concerned with the degree and extent to which the conservation practices are actually implemented. The monitoring strategy is devised during the analysis phase, since the level of implementation must be evaluated if the risk-reduction objectives are to be met. The actual monitoring is carried out during the implementation phase.

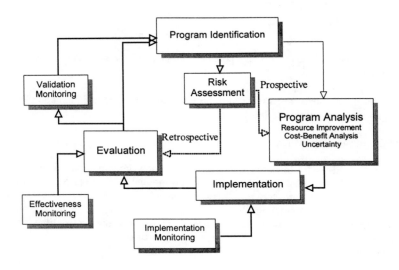

Figure 1. Overview of Monitoring and Adaptive Risk Analysis

Effectiveness monitoring is used to help characterize the degree to which the program implementation is actually providing the expected reductions in environmental risks that were projected during analysis. Program effectiveness is part of the retrospective risk assessment which characterizes the actual risk reductions achieved by the program. Monitoring of measurement endpoints is used to quantify effects on the assessment endpoints identified in the conceptual model developed during program identification.

Validation monitoring is included in the evaluation phase of the risk analysis to assess the actual risk reduction with respect to program goals and the practices implemented. It makes use of the risk characterization from the retrospective risk assessment. The purpose is to determine the effectiveness of the endpoints and measurements selected in implementing the program and reducing the level of uncertainty in making risk-management decisions. How useful were the program assumptions about risk and the implementation practices, and what are the likely management alterations necessary for greater effectiveness?

NATIONAL:

- What are the criteria for evaluation? How is successful implementation defined?

- Do the assessment endpoints correspond to management objectives? Are they ecologically relevant?

- Does the baseline reflect natural variability?

- How will the information from monitoring activities be used to evaluate the extent to which ecological risk-reduction benefits and other program objectives are being achieved?

- What kinds of provisions have been made for utilizing monitoring and evaluation results in the program decision-making process?

STATE:

- Does the national plan for monitoring and evaluation reflect regional ecosystem differences?

- Can the regional monitoring and evaluation results be used to improve state and local implementation activities?

LOCAL:

- Does the monitoring plan provide the best means for capturing the results of conservation plans?

- Are there significant ecological, economic, and social changes resulting from the program that are not reflected in the monitoring plan?

Conclusions

USDA resource conservation programs are complex and are modified over time to address multiple objectives in a broad range of ecosystems. An approach which integrates risk assessment, risk management, and monitoring and evaluation can improve the effectiveness of these programs. The adaptive risk analysis method stresses the need to establish linkages between program objectives, risk characterization, assessment endpoints, ecosystem characterization, risk mitigation, and monitoring as a means to improve program performance in reducing ecological risks. The value of the adaptive risk analysis method is a function of the extent to which the best, most reasonably available information is used in the analysis, as well as the degree to which it is an integral part of the agency's decision-making process.

The adaptivity of the process is a result of monitoring assessment endpoints which are ecologically significant and reflect attainment of program objectives. Providing opportunities for feedback among the levels of the program implementation hierarchy is also vital to the effectiveness of the adaptive risk analysis method.

References

Cleaves, D.A. 1995. Assessing and communicating uncertainty in decision support systems: lessons from an ecosystem policy analysis. *AI Applications* 3(3).

Commission on Risk Assessment and Risk Management. 1997. *Risk Assessment and Risk Management in Regulatory Decision Making*. Environmental Protection Agency, Washington, DC.

Kopp, R., A. Krupnick, and M. Toman. 1996. *Cost-Benefits Analysis and Regulatory Reform*. Resources for the Future, Washington, DC.

OMB. 1996. *Economic Analysis of Federal Regulations under Executive Order 12866*. Office of Management and Budget, Washington, DC.

USDA. 1996. *National Planning Procedures Handbook*. Natural Resources Conservation Service, US Department of Agriculture, Washington, DC.

USDA. 1997a. *Conservation Reserve Program: Environmental Risk Assessment*. Farm Service Agency, US Department of Agriculture, Washington, DC.

USDA. 1997b. *Environmental Quality Incentives Program: Environmental Risk Assessment*. Natural Resource Conservation Service, US Department of Agriculture, Washington, DC.

USEPA. 1997. *Draft Final Guidelines for Ecological Risk Assessment*. Risk Assessment Forum, US Environmental Protection Agency, Washington, DC.

Public Perceptions of Water Pollution Threats and Experts as Mitigators

Robert E. O'Connor,[1] Richard J. Bord,[2] and Ann Fisher[3]

Abstract

Most people are unsure of whether water pollution poses a threat to their personal health, but they are convinced that it poses a threat of harmful long-term impacts to society. They are uncertain of whether individuals themselves can take steps to reduce threats from water pollution, but a majority think it is likely that experts can find ways to significantly reduce those threats. This optimistic view of experts relates most strongly to a positive view of government, altruistic environmentalism, self-assessed informedness, and the belief that water pollution is a societal risk. Those who are most convinced that water pollution poses long-term risks to society are also the most likely to trust in the abilities of experts to reduce threats. In effect, the "see-no-problem" people do not think experts can reduce threats, perhaps in part because threats are not serious to them.

Introduction

Public support for risk-based decision making would seem logically to depend on a broad public belief that scientists and other experts are capable of reducing threats to society. If experts cannot make the world a safer place, why should the public trust them to vary from traditional standards-based practices? We recognize that the psychometric approach to risk perceptions has increased our

---

[1]Associate Professor, Department of Political Science, Pennsylvania State University, University Park, PA 16802
[2]Associate Professor, Department of Sociology, Pennsylvania State University, University Park, PA 16802
[3]Senior Research Associate, Department of Agricultural Economics and Rural Sociology, Pennsylvania State University, University Park, PA 16802

understanding of why people fear some risks more than others, and why expert opinion frequently differs from attitudes held by the general public. Generally absent from research in the psychometric tradition, however, is attention to opinions about government funding and the abilities of experts and individuals themselves to reduce risks to persons and to society. We know a lot about how people rate risks, but much less about what people think about the effectiveness of experts, governments, and individuals themselves to reduce risks. To broaden our understanding of risk-mitigation judgments, we explore the following questions focusing on water pollution, but in the context of other risks:

- Do people think that scientists and other experts can find ways to reduce risks to society?
- Do people think that they, as individuals, can take actions personally to reduce risks to themselves?
- Do people think that an additional $100 million in government spending directed at each problem would be helpful?

Hereafter we refer to "threats" rather than "risks" for three reasons: our questionnaire asks about threats rather than risks; we find convincing the argument of Paul Slovic (1996) that "risk" is a hopelessly muddled term; and most important, the dictionary definition of "threat" as "an indication of impending danger or harm" seems best to capture the meaning of what we are trying to measure.

After reporting how people rate the above potential mitigators, we examine correlates of public support for experts. We look to cultural, demographic, and attitudinal variables in order to explore why some people are hopeful and others pessimistic about the threat-reduction potential of scientists and other experts. These independent variables fall into two categories:

- *Attributes*, including cultural predispositions, gender, and age. These attributes are somewhat fundamental and stable for individuals.
- *Attitudes*, including the proper role of government, conservative/liberal political leanings, environmental concerns as indexed by the New Environmental Paradigm scale plus items indicating global and generational concerns, self-assessed informedness, and perceptions of the severity of both personal and societal threat posed by environmental and health problems.

Previous Research

Although tangential to the question of rating threat mitigators, several scholars have suggested variables significant to an understanding of the social construction of risk perceptions. These variables involve cultural, political, and environmental values as well as demographics.

Working initially with anthropologist Mary Douglas, Aaron Wildavsky and his collaborators argued for attention to culture (Douglas and Wildavsky 1982; Wildavsky 1987; Thompson, Ellis, and Wildavsky 1990; Wildavsky and Dake 1990). They posited that attitudes toward risk and the institutions that manage risk in our society are a consequence of people's cultural assumptions. Empirical culture studies (e.g., Dake 1991; 1992) that use the Douglas-Wildavsky formulation offer evidence, disjointed as it may be, that cultural assumptions relate to attitudes toward risks as well as to the estimated competence of institutions to ensure public health and safety. This culture theory posits four dimensions: egalitarianism, fatalism, hierarchy, and individualism.

People who score high on the *egalitarianism* scale believe that America suffers from powerful, huge organizations that exploit nature and the work force, stultifying creativity and cooperation. Egalitarians do not trust the work of large institutions, either private ones or in the public sector, and so they are wary of claims that health risks are minimal. We would expect that people who score highly on the egalitarian scale would have little faith in experts because experts usually represent large institutions. Egalitarians seemingly would think that they themselves could reduce risks by taking action in their communities.

Cultural theorist Richard Ellis (1993) has written extensively about *fatalists*, people who have low expectations regarding their ability to influence the world around them. As a result, they are concerned about becoming victims of forces they cannot manage. One element of fatalism is randomness, which would suggest a lack of faith in experts. Yet Ellis describes another element of fatalism as a paternalistic trust in powerful authorities; in some accounts it is noted that slaves or prisoners accept the legitimacy of their masters. The relationship between fatalists and trust in experts is unclear.

People who score high on the *hierarchy* scale (Dake 1991, 1992; Ellis 1993) have faith in the ability of people in authority to manage the affairs of the world well. Unless given strong reasons to doubt, their instincts are to trust their well-being to experts who, after all, have earned their degrees and work for important organizations. Scores on the hierarchy scale should relate positively to faith in experts.

Finally, Dake (1991, 1992) and Thompson et al. (1990) write that people who score highly on the *individualism* scale have strong confidence in their own ability to manage things, including risks from environmental sources. Individualists prize competition, believing that society would be better off if people were free to rise or fall based on their own performances. We expect them to believe that individuals can protect themselves from threats, and to have little faith in experts.

Besides culture theorists, other scholars of risk perceptions have proffered additional variables as important for understanding how people view risks and our ability to reduce threats. These variables include age (MacManus 1996), gender (Davidson and Freudenberg 1996; Bord and O'Connor 1997), and two varieties of environmentalism: Dunlap's (1978) New Environmental Paradigm (support for a harmonious relationship of humans with nature rather than the traditional paradigm of human dominion over nature with unlimited growth), and Stern and Deitz's (1994) altruistic environmentalism (the view that pollution is quite harmful both here and around the world, and will cause problems for future generations. In light of its widespread use and broad validation (Pierce et al. 1987), we use the New Environmental Paradigm scale as our measure of ecological environmentalism. The Stern and Deitz measure captures a different environmental dimension: concerns that environmental problems may be harming other people and future generations.

We include two political variables because so many experts are government employees or depend upon government funding to support their work. One scale measures a general belief that government is beneficial; the other is the traditional conservative/liberal self-identification. We expect these variables to correlate with trust in experts to reduce water pollution.

Finally, we include self-assessed informedness and the severity of threat perceptions. Do people who think they are informed have greater trust in the capabilities of experts? Are people who see water pollution as a serious societal problem more or less sanguine about experts?

In summary, the research reports how people rate three threat mitigators: experts, individuals themselves, and increased government spending. Then we explore the ability of variables that have accounted for variance in risk perceptions in prior research to help us understand who has faith in experts to reduce threats from environmental problems.

Participants in the Survey

In the spring and summer of 1997, 1,225 adults (18 and over) returned completed questionnaires, a response rate of 59 percent. The purchased sample mailing list represents a random sample of residential addresses from the 48 contiguous states. In comparison with census population figures, our sample over-represents males (62 percent) and persons 66 and older (24 percent). Weighting procedures produce only minimal changes in the tables for this paper, so we have not weighted the results.

Asked to participate in a study of public priorities for goals and issues affecting their communities, respondents answered five pages of questions about goals and comparative threat perceptions, four pages about climate change, four pages about their social and political values, and two pages of demographics. The purpose of the overall project is to develop and test a risk-estimation model for climate change.

<u>Measures</u>

"Water pollution" is an item in a list that includes violent crime, hazardous chemical waste, AIDS, air pollution, cancer, global warming, heart disease, and automobile accidents. The measures of perceptions of the threat posed by water pollution and of the likelihood that solutions to water-pollution threats will improve the situation are straightforward:

- Threats to my health: (In your judgment, how likely are <u>you</u>, sometime during your life, to experience serious threats to your health or overall well-being as a result of each of the following?)

- Threats to society: (In your judgment, how likely is it that each of the following will have extremely harmful, long-term impacts on our society?)

- Personal control: (In your judgment, how much can you, as an individual, personally do to reduce threats to yourself from each of the following?)

- Experts can help: (In your judgment, how likely is it that scientists and other experts can find ways to significantly reduce threats to society from each of the following?)

- Money can help: (Some people think that spending more on a specific issue can help a lot, but others think that spending more would not improve the situation. Suppose government allocates an additional 100 million dollars to EACH of these issues. In your judgment, how much would another $100 million help for each issue?)

In measuring optimism that experts can reduce threats, we could have used the simple five-point scale from the question that applies only to water pollution, despite the narrow range of this dependent variable for Table 1. Fortunately, answers to questions about experts' capabilities for different environmental and health problems scale (Cronbach's alpha = .82), so we are able to create a new dependent variable with a broader range (scores from 7 to 35). The environmental and health items that comprise the new variable are hazardous chemical wastes, AIDS, air pollution, cancer, global warming, heart disease, and water pollution.

The greatest measurement challenge is to capture the cultural dimensions. Fortunately, the Institute for Policy Analysis of the University of New Mexico (Gastil, Jenkins-Smith, and Silva 1996) extensively analyzed cultural bias items and recommended sets of questions. Following their advice and our own extensive pretesting, we used four scales:

- *Egalitarian*
  What our society needs is a fairness revolution to make the distribution of good more equal; society works best if power is shared equally.
- *Fatalist*
  Most of the important things that take place in life happen by random chance; no matter how hard we try, the course of our lives is largely determined by forces beyond our control.
- *Hierarchist*
  Society would be much better off if we imposed strict and swift punishment on those who break the rules; our society is in trouble because we don't obey those in authority.
- *Individualist*
  Even the disadvantaged should have to make their own way in the world; people who get rich in business have a right to keep and enjoy their wealth; even if some people are at a disadvantage, it is best for society to let people succeed or fail on their own.

These scales emerged from the application of factor analysis and reliability scaling. They – and the other scales reported in this study – have a Cronbach's Alpha of at least .60, and each item's factor loading is at .60 or higher.

The short version of Dunlap's New Environmental Paradigm (1978) has six items:

1) the earth is like a spaceship with only limited room and resources;
2) plants and animals do *not* exist primarily to be used by humans;
3) modifying the environment for human use seldom causes serious problems (disagree);
4) there are no limits to growth for nations like the United States (disagree);
5) the balance of nature is very delicate and easily upset by human activities;
6) mankind was created to rule over the rest of nature (disagree).

Although Stern, Deitz, and Kalof (1993) use a three-item scale for altruistic environmentalism, the failure of one of their items to scale adequately in our analysis induced us to use only two: "the effects of pollution on public health are worse than we realize," and "pollution generated in the United States harms people world-wide."

The scale for government effectiveness derives from three items for which respondents selected one of two statements:

1) Government is almost always wasteful and inefficient.   OR
   Government often does a better job than people give it credit for (selected).

2) Government regulation of business is necessary to protect the public interest (selected).   OR
Government regulation of business usually does more harm than good.

3) Most elected officials care what people like me think (selected).   OR
Most elected officials don't care what people like me think.

Scores range from "3" (government does better job than given credit for, regulation necessary, and elected officials care) to "0" (government wasteful, regulation harmful, officials don't care).

The liberalism measure is a single 5-point item that asks respondents to rate themselves as very conservative, conservative, neither, liberal, or very liberal.

The question of whether "experts can find ways to significantly reduce threats" focuses on their ability to reduce threats to society, not to individuals. Therefore, in Equation (2) of Table 1, the independent variables that measure the severity of threats also use questions that focus on threats to society and not to individuals. The two scales used for "environmental threats severe" and "health threats severe" derive from the question, "In your judgment, how likely is it that each of the following will have extremely harmful, long-term impacts on our society?" "Environmental threats severe" includes hazardous chemical waste, air pollution, global warming, and water pollution. "Health threats severe" includes cancer and heart disease.

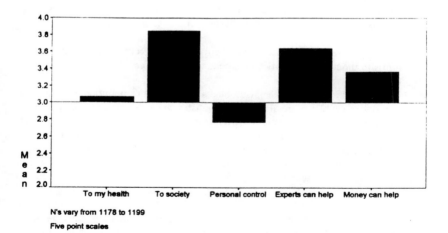

N's vary from 1178 to 1199

Five point scales

Figure 1. Water Pollution Threats

Results

Figure 1 presents the results of five questions regarding threats from water pollution. Most respondents are ambivalent about whether water pollution is likely to cause them health problems in their lifetimes. The mode is "3," the middle category of the five-point scale. When the focus shifts to the threat to society, most respondents think that a long-term harmful impact is likely.

Regarding what can be done to mitigate threats from water pollution, there is ambivalence about whether individuals can take steps personally to reduce risks to themselves. The mode is "3," the middle category. More optimistically, a plurality thinks that increasing government spending on water pollution by $100 million would help a lot, although the mode again is "3." Most optimistically, a majority think that it is likely that experts can significantly reduce threats from water pollution. Also, there is little polarization as few people check the "1" or "5" extremes for any option. What stands out is the majority support for the likelihood that experts can significantly reduce threats.

Table 1 reports the two equations that account for variance in support for experts. There are four notable results:

- The cultural and demographic variables account for only 2 percent of the variance by themselves, and lose all statistical significance when environmental and political attitudes enter the equation. Egalitarians and women are slightly more likely to support experts, but these variables lose independent explanatory power when we add the attitudinal variables to the equation. Trust in experts to reduce threats is not tied to any cultural predisposition, nor to any age group or sex.

- People who are more positive about government effectiveness are also more optimistic about the capabilities of experts to reduce threats.

- People who think environmental problems such as water pollution are likely to cause severe problems are also more optimistic about the capabilities of experts to reduce threats. "Altruistic environmentalism," essentially a measure that pollution elsewhere is worse than we think, and "severity of environmental threats to society" both measure the seriousness of environmental problems. Far from feeling that we are at risk because of the incompetence of experts, people who think society faces a difficult future are more likely to expect significant advances from scientists and other experts.

- People who think they are well-informed about environmental problems are more likely to think that experts can reduce threats from those problems. In the absence

Table l. Effects of Culture, Gender, Age, Environmental and Political
Attitudes, Knowledge, and Perception of Severity of Danger on
Perceptions of the Efficacy of Experts in Reducing Environmental and
Health Threats

	1	2
Egalitarian	.26**(.09)	.07 (.09)
Fatalist	-.10 (.09)	-.05 (.10)
Hierarchist	-.06 (.09)	-.05 (.10)
Individualist	-.14 (.08)	.02 (.08)
Female gender	.82* (.37)	.59 (.39)
Age	-.11 (.12)	-.15 (.12)
New environmental paradigm		.03 (.05)
Altruistic environmentalism		.25** (.11)
Government effectiveness		.42*** (.09)
Liberalism		.02 (.20)
Environmental threats severe		.18** (.06)
Health threats severe		.11 (.10)
Informed		.21*** (.06)
Adjusted R^2	0.02	0.08
N	1035	916

Cell entries are unstandardized regression coefficients, with standard errors in parentheses.
*   Significant at .05
** Significant at .01
*** Significant at .001, all two-tailed tests

of data over time, we do not know causal connections. Some people may feel that they need to become informed because they have heard that environmental problems are serious. In becoming informed, they may develop respect for experts. Other people may decide to become informed for reasons of general curiosity. As they learn, they may then conclude that environmental problems are serious, but that experts are likely to reduce threats.

Conclusions and Discussion

At first glance, it is hard to generate enthusiasm for findings that show moderation in public opinion. A presentation about a public demanding certain risk-mitigation approaches, or prepared to revolt to stop experts from doing their research, would be more interesting than our findings. Indeed, a listing of what we did *not* find highlights the significance of our actual findings.

First, we did *not* find an American public polarized between one group that fervently rejects the efficacy of any potential mitigator and a second group that is certain that scientists or governments or individuals themselves can reduce threats to society and to individuals from water pollution. Such a finding would complicate the task of policy-makers who would have to explain and justify their decisions to a populace with strong views. Instead, most people are in the middle, slightly optimistic or slightly pessimistic that these mitigation approaches can work. Mushiness here signals an opportunity for public education and risk communication.

Second, we did *not* find a cynical public united in an anti-expert consensus. More people agree rather than disagree that scientists and other experts can find ways to significantly reduce threats from water pollution. The support may be tepid, but tepid support provides the sort of permissive public opinion that allows scientists to influence public policy (Sabatier and Jenkins-Smith 1993).

Third, we did *not* find that variables of demography, culture, environmental values, political opinions, and judgments of the severity of the problems predict most of the variance in faith in experts. Opinions about the ability of experts to reduce threats correlate with altruistic environmentalism, government effectiveness, self-assessed informedness, and threat severity. Yet, all of the social and opinion variables together account for under 10 percent of the overall variance. Discussions over expert proposals are not likely to be burdened heavily with ideological cleavages, with everyone falling on one side or the other, conditioned by their demographics, cultural attitudes, and environmental and political opinions. When people are discussing how experts can reduce threats, they can *really* be talking about strategies to reduce threats, not lined up with the young against the old, environmentalists against developmentalists, conservatives against liberals, and egalitarians against hierarchists. The R^2s we report indicate that people with the same demographics, cultural assumptions, and attitudes often rate the ability of experts to reduce threats quite differently. Survey researchers may find it more satisfying to report high correlation coefficients with such variables, but high R^2s would represent cultural and attitudinal cleavages that could hinder constructive debate about how best to use experts to reduce threats.

Acknowledgments

We gratefully acknowledge financial support from the National Science Foundation (Grant SRB-9409548) and the US Environmental Protection Agency (Cooperative Agreement CR 824369-01). The views expressed are the authors', and should not be attributed to their employer or funding sources. We also appreciate the willingness of Hank Jenkins-Smith to share the cultural item analysis. We thank Craig Ortsey and Andrew Shults for outstanding research assistance.

References

Bord, R.J., and R.E. O'Connor. 1997. The gender gap in environmental attitudes: the case of perceived vulnerability to risk. *Social Science Quarterly* 78(4): 830-840.

Dake, K. 1991. Orienting dispositions in the perception of risk: an analysis of contemporary worldviews and cultural biases. *Journal of Cross-Cultural Psychology* 22(1): 61-82.

Dake, K. 1992. Myths of nature: culture and the social construction of risk. *Journal of Social Issues* 48(4): 21-37.

Davidson, D., and W. Freudenburg. 1996. Gender and environmental risk concerns: a review and analysis of available research. *Environment and Behavior* 28: 302-39.

Douglas, M., and A. Wildavsky. 1982. *Risks and Culture: An Essay on the Selection of Technological and Environmental Dangers*. Berkeley, CA: University of California Press.

Dunlap, R. E. 1978. The new environmental paradigm: a proposed measuring instrument and preliminary results. *Journal of Environmental Education* 9: 10-19.

Ellis, R. 1993. *American Political Cultures*. Oxford, UK: Oxford University Press.

Gastil, J., H. Jenkins-Smith, and C. Silva. 1996. *Analysis of Cultural Bias Survey Items*. Typescript, Institute for Public Policy, University of New Mexico, Albuquerque, NM.

MacManus, S. A. 1996. *Young v. Old: Generational Combat in the 21st Century*. Boulder, CO: Westview Press.

Pierce, J.C., P.N. Lovrich, and T. Tsurutani. 1987. Culture, politics, and mass publics: traditional and modern supporters of the new environmental paradigm in Japan and the United States. *Journal of Politics* 49(1): 54-79.

Sabatier, P., and H. Jenkins-Smith (Eds.) 1993. *Policy Change and Learning: An Advocacy Coalition Approach*. Boulder, CO: Westview Press.

Slovic, P. 1996. Trust, emotion, sex, politics, and science: surveying the risk-assessment battlefield. In *Environment, Ethics, and Behavior*. M. Bazerman, D. Messick, A. Tenbrunsel, and K. Wade-Benzoni (Eds.). San Francisco, CA: Jossey-Bass.

Stern, P., T. Dietz, and L. Kalof. 1993. The value basis of environmental concern. *Journal of Social Issues* 50(3): 65-84.

Thompson, M., R. Ellis, and A.Wildavsky. 1990. *Cultural Theory*. Boulder, CO: Westview Press.

Wildavsky, A. 1987. Choosing preferences by constructing institutions: a cultural theory of preference formation. *American Political Science Review* 81(1): 2-22.

Wildavsky, A., and K. Dake. 1990. Theories of risk perception: who fears what and why? *Daedalus* 119(4): 41-60.

Time-Dependent Reliability and Hazard Function Development for Navigation Structures in the Ohio River Mainstem System Study

Robert C. Patev[1], MASCE, Bruce Riley[2], MASCE,
David Schaaf[3], MASCE, and Nathan Kathir[4], MASCE

Abstract

This paper describes the time-dependent reliability methodology and hazard function development for navigation structures in the Ohio River Mainstem System Study (ORMSS). The ORMSS is currently examining reliability over the next 50 years for structural, mechanical, and electrical components at 19 lock and dam structures. The time-dependent approach discussed in this paper accounts for the cumulative effects of degradation on reliability. Hazard rates for the components are developed based on the concept of "life cycles". The implementation of these methods is handled using Monte Carlo simulation for a 50-year life-cycle period. Illustrative examples are highlighted for a vertical beam component of a miter gate and for an anchored sheet-pile wall.

Introduction

The US Army Corps of Engineers, through the Louisville, Pittsburgh, and Huntington Districts, is currently performing the Ohio River Mainstem System Study (ORMSS). The objectives of the ORMSS are to develop the major investment strategies for required capital improvements to the Ohio River navigation system over the next century. The ORMSS navigation region consists of 19 locks and dams that stretch over 981 river miles from Pittsburgh, Pennsylvania, to Cairo, Illinois.

[1]Civil Engineer, USAE Waterways Experiment Station, 3909 Halls Ferry Road, Vicksburg, MS 39180
[2]Civil Engineer, Headquarters, US Army Corps of Engineers, 20 Massachusetts Avenue NW, Washington, DC 20314
[3]Civil Engineer, USAE Louisville District, 600 Martin Luther King Jr. Place, Louisville, KY 40202
[4]Civil Engineer, USAE St. Paul District, 180 Kellogg Blvd., St. Paul, MN 55101

This ORMSS navigation area is in full operation year-round and has in excess of 230 million tons of cargo that are shipped annually.

ORMSS has focused the engineering portion of the study on determining the reliability for the major components of these locks and dams. Some of the navigation structures in ORMSS were built as far back as the 1920s and have suffered significant deterioration and degradation over time. The degradation rates and time-dependent reliability models must be properly developed to predict future performance and reliability. Most importantly, these same reliability models must be calibrated to the past field performances of the ORMSS structures.

Time-Dependent Reliability Analysis

Numerous methods are found in the literature to determine the time-dependent reliability for engineering components. Putcha and Patev (1997) summarize over 40 different available methodologies. The primary goals of the ORMSS engineering reliability efforts were not to reinvent the concept of time-dependent reliability, but to select, develop, and implement usable methods for its engineering structures. In addition, these reliability values are utilized not to make safety decisions, but to make water resource investment decisions for the ORMSS.

Previous reliability work by the US Army Corps of Engineers (Headquarters, Department of the Army 1992) has focused on reliability index, or β methods. These give a good estimate of the relative reliability and are easy for engineers to implement within a spreadsheet. A difficulty with using these methods, however, is that they give only an instantaneous snapshot in time. Hence, they do not reflect the cumulative behavior of degradation in a structure.

The time-dependent reliability method used for the ORMSS models needs to incorporate the time relationships of strength R (or capacity), vs. the load S (or demand). With structural degradation, the demand (load) at some point in time will exceed the capacity (strength) which has degraded with time. Figure 1 shows this time-dependent relationship of strength vs. load. Unsatisfactory performance will occur when load exceeds strength. This is the basis for the concept of a "life cycle" (to be discussed later) which utilizes the development of a probability density function (pdf) that reflects the range of uncertainties in both strength and load over time.

The limit state for strength, $R(t)$, and load, $S(t)$, can be expressed as follows

$$R(t) - S(t) < 0 \tag{1}$$

$$Pu(t) = P[R(t) < S(t)] \tag{2}$$

where $Pu(t)$ is the probability of unsatisfactory performance (PUP) at time t.

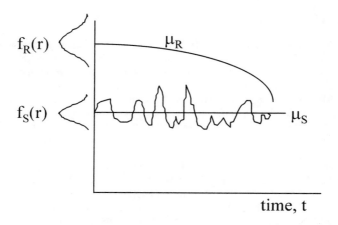

Figure 1. Time-Dependent Relationship of Strength vs. Load

However, Ellingwood (1995) points out that this reflects only a snapshot of the reliability of a structure in time. It does not indicate the likelihood that the structure fails to survive the interval (t_1, t_2). In addition, he states that $Pu(0,t)$ is not equivalent to $P[R(t0<S(t)]$ since it does not account for any cumulative effects in performance. Utilizing these concepts and methodology, Ellingwood's time-dependent reliability methods were implemented in the ORMSS.

Ellingwood and Mori (1993) have established the equation for time-dependent reliability, $L(t)$ (L instead of R because R was already assigned to strength) as:

$$L(t) = \int_0^\infty \exp[-\lambda t[1 - 1/t \int_0^t F_S(g(t)r)dt]]f_R(r)dr$$

where

F_S = CDF of Load   (3)

g(t)r = time - dependent degradation

$f_R(r)dr$ = pdf of initial strength

λ = mean rate of occurrence of loading

However, this equation is not implemented easily without making certain assumptions, and closed-form solutions are only available for special cases. The solution to this problem was to utilize Monte Carlo simulations to examine the "life cycle" for a component or structure.

Life-Cycle Concept

Field data is simply not available to establish the frequency of unsatisfactory performances for ORMSS navigation structures. This is because unsatisfactory performances in ORMSS structures have been completely eliminated through careful design. Since the field data does not exist, a "life-cycle" concept was developed by the ORMSS to assist in defining the probability density functions (pdf) and cumulative density functions (CDF) for specific performance modes in the life of a structure. The term "life cycle" was employed since this methodology is similar to that used in developing the annual costs over the life of the structure.

This life-cycle concept establishes the pdf and CDF for specific component limit states. A flow chart defining the steps of a life cycle is shown in Figure 2. The process utilizes Monte Carlo simulation to develop the pdfs for both strength, $R(t)$, and load, $S(t)$, over time, as shown in Figure 1 and discussed above. These variables of strength and load are then propagated down a complete life cycle. A typical example would be 50 years for a civil structure. The limit state for a mode of performance is examined for the life cycle, and the year in which unsatisfactory performance occurs is recorded. A new life cycle is then initialized with a new set of random variables for strength and load. This process is iterated until the reliability over time converges to a specified tolerance.

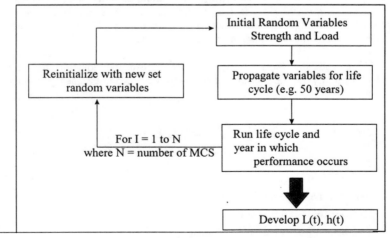

Figure 2. Flow Chart of Life-Cycle Simulations

Hazard Function Development

The life-cycle method described above develops the density functions for the time-dependent reliability of specific performance modes. Hazard functions or hazard rates are also developed to assist the ORMSS economists with the overall system analysis to determine their water-resource investment strategy. Frequently, hazard functions also are termed in the literature as failure rates (instantaneous or conditional), force of mortality, or mortality rates. These definitions can be found in many engineering reliability textbooks (e.g., Vinogradov 1991; O'Connor 1985; Henley and Kumamoto 1985) or in various handbooks on life data analysis (e.g., Nelson 1981).

The conditional hazard rate $h(t)$ used in the models for the ORMSS can be defined as:

$$h(t) = P[failin(t, t + dt)|survived(0,t)]$$

or

$$h(t) \approx -\frac{d(\ln L(t))}{dt} \tag{4}$$

or

$$= \frac{f(t)}{L(t)} = \frac{1 - F(t)}{L(t)}$$

where $L(t)$ = cumulative time-dependent reliability, $f(t)$ is the pdf, and $F(t)$ is the CDF.

Unfortunately, this type of hazard function methodology requires addressing certain simulation issues to maintain consistent results. Strict guidelines, standards, and set procedures need to be followed to insure the proper review of the results. These guidelines for navigation structures in ORMSS have been developed by Patev, Riley, Schaaf, and Kathir (1998). A critical issue is the convergence of the simulation results. Convergence of the reliability needs to be carefully checked not only for the performance limit state, but for the reliability for each year of the life cycle. The convergence will dictate the number of simulations that need to be executed. For ORMSS models, typically this has ranged from about 10,000 to 50,000 simulations.

Another issue is the sensitivity of the models to the input random variables. If the limit state is highly sensitive to a specific random variable and there is very little confidence in its statistical parameters, a new examination of the random variable is required or the results from the model can be seriously questioned. Lastly, and most important to the ORMSS, is calibrating the reliability model to structures in service. This gives a true validation of the model and the required input random variables and their distributions. In turn, the validation produces a level of confidence in estimating the reliability of the structure into the future.

Illustrative Example – Vertical Beam of a Miter Gate

This example investigates the performance of a vertical beam in a miter gate for a 50-year life-cycle period. One of the prime degradation mechanisms is corrosion of the beam in the splash zone due to cyclic daily fluctuations in the lower pool elevation. Another important degradation mechanism is fatigue in the beam due to a cyclic head differential caused by the transit of barge traffic. This mechanism is not covered in this paper but is discussed in Patev, Riley, Schaaf, and Kathir (1998). Figure 3 shows a diagram of the loading on a vertical beam due to the differential pool, h_i, created by an upper pool, h_u, and lower pool, h_l.

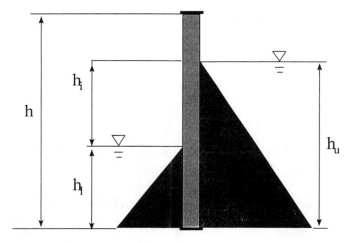

Figure 3. Loading on Vertical Beam of Miter Gate

The capacity of the vertical beam is derived primarily from the yield strength, f_y, of the steel beam. f_y is a random variable that can be determined from laboratory test data. The degradation of the structure occurs from corrosion of the beam in the splash zone. This corrosion causes a decrease in section modulus, $S(t)$, over time. The degradation rate can be determined by using the power law for corrosion, which is represented by

$$C(t) = At^B$$

where

A is a random variable from measured field data

B is deterministic and determined from measured field data

(5)

The demand on the vertical beam comes from the loading of the differential pool, h_i. The differential pool is based on the differences between the upper and lower pools. This equation can be expressed as

$$h_i = h_u - h_l$$
where h_i = differential pool
h_u = *upper* pool   (6)
h_l = *lower* pool

The upper pool for a control dam structure can be treated as a constant. For a fixed dam structure, the values can vary depending upon flow conditions. Lower pool elevations are not a constant due to variations in the hydraulic regime and lack of flow control. The annual maximum lower pool elevations are taken as random variables that can be determined from years of daily hydraulic data. The statistics for pool distributions for the ORMSS locks and dams can be found in Patev (1998).

The limit state for the vertical beam is based on the moment of the beam within the splash zone. Note that this limit state is not based on failure of the beam (i.e., collapse), but on the demand exceedance in the moment capacity for the beam. The equation for the ultimate limit state for a vertical beam can be expressed as

$$M_{capacity} - M_{demand} < 0 \qquad (7)$$

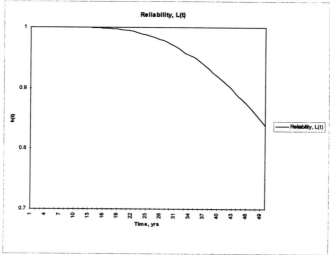

Figure 4. Time-Dependent Reliability of Vertical Beam of Miter Gate

The results for time-dependent reliability and the conditional hazard rate using the life-cycle concepts described in the preceding paragraphs are shown in Figures 4 and 5. The reliability at the time of 50 years is 0.84 and the hazard rate is 0.017 or 1.7%. This conditional hazard rate means that the beam has a 1.7% chance of exceeding the moment in year 50 if it has survived up to year 49.

In addition, attention should be paid to the shape of the hazard curve in Figure 5. The initial portion of the bathtub curve, or the early failure region that is characteristic of the bathtub, is absent. This is because the yield strength, f_y, of the steel is guaranteed for a minimum strength of 36 ksi. In addition, this portion of the curve is removed because of the safety factor introduced in the original design of the vertical beam. For a further explanation of this example, refer to Patev, Riley, Schaaf, and Kathir (1998).

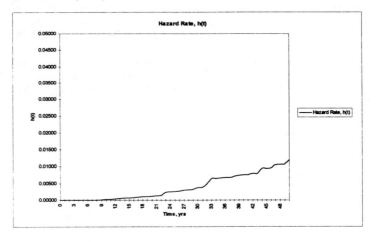

Figure 5. Hazard Rate for Vertical Beam of Miter Gate

Illustrative Example – Anchored Sheet-Pile Wall

Anchored sheet-pile walls have three typical modes of unsatisfactory performance: corrosion of the anchor, corrosion of the sheet pile, and loss of embedment depth due to scour. These modes of performance are shown in Figure 6. This example focuses on the performance mode for the capacity exceedance of the anchor rod. The other performance modes are discussed further in Patev and Riley (1998). The procedures used to analyze an anchored sheet-pile wall can be found in publications such as the *USS Steel Sheet Piling Design Manual* (US Steel Corporation 1970) and the *User's Guide for the Computer Program, CWALSHT*, (Dawkins 1991). The process invoked for this example involves developing an iterative procedure to solve for the anchor force based on both the force and moment equilibrium of the active and passive earth pressures that result from the soil and water pressures. The freebody diagram for anchor sheet-pile wall equilibrium is shown in Figure 7.

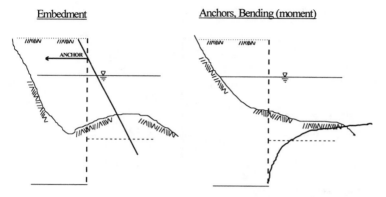

Figure 6. Performance Modes for Anchored Sheet-Pile Walls

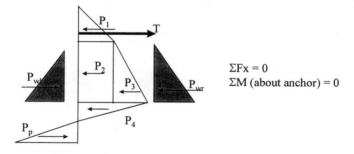

Figure 7. Freebody Diagram for Anchored Sheet-Pile Wall

The geometry for the anchored sheet-pile wall is shown in Figure 8. The wall is composed of a PZ-32 piling with an anchor-rod tieback at an elevation of 4 ft. The strength of the anchor is based on the yield strength of the steel in the anchor rod, f_y. This random variable is determined from the guaranteed minimum yield strength from laboratory tests. The degradation to the strength of the anchor is caused by the corrosion and section loss of the anchor rod. The annual rate of corrosion is a random variable determined from field measurements.

The loading in the anchor is due to the force and moment equilibrium with the soil and water pressures. This can be attributed to the active and passive pressure coefficients, K_a and K_p, which are random variables based on the soil parameters for internal friction angle ϕ and wall friction δ. The unit weight of the soil layers, γ, is also a random variable that was obtained from soil-test data.

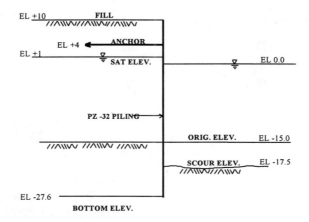

Figure 8. Geometry for Anchored Sheet-Pile Wall Example

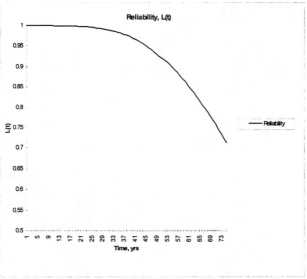

Figure 9. Reliability of Anchored Sheet-Pile Wall

The limit state for the anchor rod is based on the force exceedance in the rod. The ultimate limit-state equation can be expressed as

$$F_{capacity} - F_{demand} < 0 \tag{8}$$

The results for the reliability and hazard functions are shown in Figures 9 and 10. The reliability at year 50 is 0.72 and the corresponding hazard rate would be approximately 0.02 or 2% at year 50, given that the structure has survived up to that time. For a detailed explanation of this example, refer to Patev and Riley (1998).

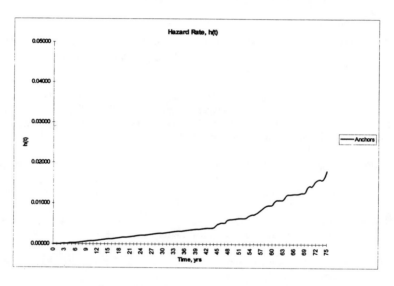

Figure 10. Hazard Rate for Anchored Sheet-Pile Wall

Conclusions

The methodologies presented in this paper can be used to develop cumulative time-dependent reliability models and hazard functions for engineering structures and their components where performance data may not be available. The concept of life cycles assists greatly with the development of hazard rates for engineering structures. Without sufficient data, hazard rates would be difficult to estimate using other procedures such as probability plotting or Weibull distribution fitting. The illustrative examples successfully document the application of these procedures for navigation structures in the ORMSS.

Acknowledgments

The authors would like to acknowledge the funding support for this work through the Ohio River Mainstem System Study. Permission was granted by the Chief of Engineers to publish this information. The views of the authors do not purport to reflect the positions of the Department of the Army or Department of Defense.

References

Dawkins, W.D. 1991. *User's Guide to the Computer Program, CWALSHT.* Technical Report ITL-91-1, US Army Engineer Waterways Experiment Station, Vicksburg, MS.

Ellingwood, B. 1995. *Engineering Reliability and Risk Analysis for Water Resource Investments: Role of Structural Degradation in Time-Dependent Reliability Analysis.* Technical Report ITL-95-3, US Army Engineer Waterways Experiment Station, Vicksburg, MS.

Ellingwood, B., and Y. Mori. 1993. Probabilistic methods for condition assessment and life prediction of concrete structures in nuclear power plants. *Nuclear Engineering and Design* 142: 155-166.

Headquarters, Department of the Army. 1992. *Reliability Assessment of Navigation Structures.* Engineering Technical Letter 1110-2-532.

Henley, E. J., and H. Kumamoto. 1985. *Reliability Engineering and Risk Assessment.* Englewood Cliffs, N.J: Prentice-Hall, Inc.

Nelson, W. 1981. *Applied Life Data Analysis.* New York, NY: John Wiley and Sons.

O'Connor, P.D.T. 1985. *Practical Reliability Engineering.* New York, NY: John Wiley and Sons.

Patev, R.C. 1988. *Load Histograms and Pool Distributions for the Ohio River Mainstem Systems Study.* Technical Report ITL-98-XX (in preparation), US Army Engineer Waterways Experiment Station, Vicksburg, MS.

Patev, R.C., and B. Riley. 1988. *Time-Dependent Reliability Analysis for Anchored Sheet-Pile Walls.* Technical Report ITL-98-XX (in preparation), US Army Engineer Waterways Experiment Station, Vicksburg, MS.

Patev, R.C., B. Riley, D. Schaaf, and N. Kathir. 1998. *Time-Dependent Reliability Analysis and Reporting Guidelines for the Ohio River Mainstem Systems Study*. Technical Report ITL-98-XX (in preparation), US Army Engineer Waterways Experiment Station, Vicksburg, MS.

Putcha, C, and R.C. Patev. 1997. Time-Variant Reliability Methods. *CRC Handbook on Uncertainty in Civil Engineering Analysis*. Boca Raton, FL: CRC Press.

US Steel Corporation. 1970. *USS Steel Sheet Piling Design Manual*. Pittsburgh, PA: US Steel Corporation.

Vinogradov, O. 1991. *Mechanical Reliability – A Designer's Approach*. New York, NY: Hemisphere Publishing Corporation.

Dam Safety and Risk of Extreme and Catastrophic Events

Wesley W. Walker,[1] Rapporteur

Introduction

Session Chair David Moser, US Army Corps of Engineers

The Corps intends to embark on an effort to develop risk-based assessments of dam safety. An issue that was once dead is now revived.

Risk-Based Decision Making for Dam Safety

Jerry Foster, US Army Corps of Engineers

The USACE has applied risk assessment in hydropower rehab studies, and is now trying to put together policy for dam safety risk analysis.

Challenges:
- Aging infrastructure
- 65% of 569 dams have hydrologic or seismic deficiencies
- Most are never subjected to the design load
- Safety modifications are expensive, the consequences of failures extreme
- Balancing risks against funding limitations
- Competing with new construction

The Corps' current analytic techniques give more weight to rare events over more likely events, downstream populations are not considered, and it doesn't do

---

[1]Regional Economist, Ohio River Region Navigation Planning Center, Huntington District, US Army Corps of Engineers, 502 Eighth Street, Huntington, WV 25701-9020

loss-of-life estimates. However, the Corps has in place a basic framework for analyzing risk; a draft engineering pamphlet will be released soon for comment.

The risk management program will:

- Use both total and individual risk to make investment decisions
- Quantify budget decisions' impacts on public safety
- Use economics to compare alternatives with equal risk
- Establish acceptable level of risk – (the Corps tries to avoid placing value on human life; limit total project risk to one fatality over project life)
- Prioritize investments on the basis of risk

Foster outlined future research and development efforts. These include:

- Developing methods and tools
- Delimiting acceptable levels of risk
- Defining the frequency of a rare event
- Describing failure modes
- Selecting target probabilities of failure

Dam Safety Evaluation and Decision Making in Australia

Terry Glover, Utah State University

In Australia, there has been an interest in risk analysis/assessment for some time. Dam safety concerns have developed in the context of the semi-privatization of water resource projects.

Several questions are asked in evaluating safety factors. Factors used to evaluate safety are:

- Is a dam state-of-the-art?
- What is the earthquake threat?
- What is the age of the project?
- What is the downstream population?

The debate continues on the relative merits of a standards-based approach vs. the risk-based approach in cost effectiveness. The standards-based approach does not specifically recognize the trade-off between risks and costs. Performance is measured based upon compliance with prescribed guidelines and/or current state-of-the-art practices. Risk-based approaches, on the other hand, rely upon engineering assessments and the identification and measurement of consequences from unsatisfactory performance.

Standards and risk-based approaches are not necessarily incompatible. Risk assessment forces you into a rolling-standards approach. A standards-based approach yields just one point on a continuum.

Cost-effectiveness measures have also been suggested. The goal in using such measures is to minimize the cost per statistical life saved plus remediation costs.

Risk-Based Dam Safety Evaluation and Improvements

Dave Cattanach, BC Hydro

Cattanach argues that we have to change from standards-based to risk-based because not all dams are the same and we want to avoid unnecessary costs. He suggests that risk, as defined below, should not exceed pre-defined criteria.

Risk = (Probability of failure) x Consequence

An example of such criteria might be: loss of life .001 lives/year and $10,000/yr. Completing this type of analysis requires an accurate assessment of the consequences of failure. Initially, assessments relied on judgmental probabilities and statistical representations in order to overcome the lack of data. Risk assessment now involves screening-level assessments in order to identify important variables, followed by detailed assessments.

BC Hydro is focusing on these kinds of variables:
- Seismic input
- Probability of liquefaction
- Deformation given liquefaction
- Failure due to liquefaction
- Consequence of failure

BC Hydro is currently working with utility companies in Australia, New Zealand, and the United States to improve the process. The focus is on:
- Improved hydrologic input – modeling extreme-event rainfalls
- Improved seismic input
- Improved consequence evaluation. Example: seasonal expected loss of life

We no longer say that a dam meets international standards; we now say it meets tolerable risks criteria. Cattanach suggests that it will take a lot of work to get risk-based tools up to the level of standards-based tools.

Achieving Public Protection in Dam Safety Decision Making.

Chuck Hennig, US Bureau of Reclamation (USBR)

The US Bureau of Reclamation has 382 high and significant dams, but many are old, built during the heydays of 1940-1970. The USBR no longer builds dams; they are in a position of assessing what can be done vs. what the dam was designed to do.

The Bureau's risk management efforts are centered around:
- monitoring structure performance,
- developing emergency action plans,
- completing aggressive dam inspections, and
- training dam tenders.

The Bureau is integrating risk assessment techniques to better evaluate uncertainties. Risk is defined in terms of annual loss of life. This is arrived at as follows:

$$\text{Prob. of load x Prob. of failure x Prob. of exposure x Consequence} = \text{Risk}$$
$$\text{Risk} = \text{Annual Prob. x Loss of life}$$
$$\text{Risk} = \text{Annual loss of life}$$

Hennig listed the Bureau's dam safety risk assessment objectives as:
1. recognizing that all dams have a risk of failure,
2. considering all factors that contribute to risk,
3. identifying the most significant factors influencing risk and uncertainty,
4. identifying alternatives,
5. setting appropriate budgets and priorities, and
6. gaining credibility and stakeholder consensus.

There are no definitive sets of guidelines for acceptable risk; however, the Bureau uses a two-tier system:
- *Tier 1*: When risk reaches certain levels, the USBR responds. The farther you go below the .001 LOL (annual probability of loss of life) line, the less justification you have to address risk.
- *Tier 2*: The Bureau also recognizes the public's aversion to single, high-consequence events. Therefore, the Bureau has decided that maximum annual probability of failure for each load category on a dam cannot exceed .001.

It was suggested by Yacov Haimes that the Bureau ought to use the term *public protection* rather than *acceptable risk*, a term that may convey an attitude of callous indifference.

SESSION VII

Regulatory Analysis

Jonathan W. Bulkley, F.ASCE[1]
Rapporteur

Opening Remarks

Session Chair Michael Krouse, US Army Corps of Engineers

In his opening remarks, Michael Krouse observed that the purpose of this session is to examine how risk analysis fits into government regulatory programs. To accomplish this objective, we need to examine how far we have come. On the one hand, regulatory agencies have worked to achieve uniform standards which in turn have led to complex regulations. Accordingly, it is important to compare and contrast the application of risk analysis between operating agencies and regulatory agencies.

Role of Risk Analysis in Regulatory Analysis

Elizabeth L. Anderson, President, Sciences International (formerly with US EPA)

Dr. Anderson's presentation reviewed the practical applications of risk-based decision making. These applications include corporate environmental management, regulatory agencies setting standards for discharges and/or exposure to pollutants in the environment, and the use of risk-based decision making in litigation and other legal activities. In all of these areas, the application of risk assessment has been immense.

---

[1]Professor of Natural Resources, Professor of Civil and Environmental Engineering, University of Michigan, School of Natural Resources and Environment, Ann Arbor, MI 48109-1115

Dr. Anderson then traced progress over the past twenty-two years in the application of risk concepts to human health, especially as applied to cancer. She traced the evolution of risk-based techniques in helping to identify whether or not particular chemicals are carcinogenic. This evolution has altered as our knowledge of cancer has grown. Dr. Anderson summarized information on exposure pathways and emphasized the importance of site-specific data. She concluded with a very interesting presentation on key concepts of ecological risk assessment and what constitutes significant risk.

Discussion

The discussion opened with a question about site-specific data which appears to be ignored in spite of being above the default level for action. Dr. Anderson observed that the approach the US EPA has taken to characterize dose-response from exposure is very conservative. The regulatory agency's objective is to have a series of models that never underestimate the risk; under this approach, one would normally expect that site-specific locations would have lower risk levels. It may be important to involve experts from local universities to ascertain what is happening at these sites.

Another question raised the issue of whether or not the regulatory agencies consider an extreme-events approach in addressing carcinogenic chemicals. Dr. Anderson responded by observing that agencies are not considering extreme events in developing dose-response models. Uncertainties need to be made more explicit and decision makers should formally recognize the extreme-event approach.

Asked about the dimensions of significant risk, in this context Dr. Anderson defined it as the probability of getting cancer in your lifetime. The regulatory agencies are moving toward expressing this probability as a range of values rather than as a single point estimate. A final discussion took place around the issue of uncertainty associated with dose-response information. Dr. Anderson observed that it is very important to communicate this information in a more effective and comprehensive way to decision makers.

Risk Assessment and Risk Management of USDA's
Environmental Quality Incentive Program

Ronald L. Meekhof, US Department of Agriculture

Ron Meekhof reported on the progress being made by the US Department of Agriculture (USDA) to apply risk analysis and risk management to a range of agency issues. These include forest pests, food safety, the National Resource Conservation Service (to reduce environmental burdens from agricultural practices), and problems arising from the importation of animal and plant species. In his opening remarks,

Meekhof stated that risk management and risk-reduction activities need to be compatible. Risk analysis/risk management processes need to be adaptive, lead to the formation of effective management practices, and clearly demonstrate whether or not risk-reduction objectives are achieved. In order to have an effective risk assessment/risk management program, the following three issues are particularly important:

(a)   The assessment end-points should accurately reflect the risks the program is intended to mitigate.

(b)   Among the range of potential measures, the most effective mitigation measures should be identified.

(c)   The most relevant data should be collected to ensure an effective risk-management program.

The presentation focused upon the Environmental Quality Incentives Program ($200 million/yr.) currently being implemented by the USDA. This is concentrated in conservation priority areas and utilizes a decentralized decision-making process to consider multiple stresses impacting multiple ecosystems.

Meekhof emphasized three stages in the USDA approach. These include:

(a)   Program identification: objectives of the program, a decentralized nine-step activity accomplished at the site level.

(b)   Conceptual diagrams: cause/effect pathways of environmental risks; identify assessment end-points with at-risk resources; identify the most significant hazards and assessment end-points.

(c)   Analysis: EPA risk mitigation measures: plan and describe program mitigation measures and measures of program benefits; identify most effective/efficient mitigation measures; provide resource improvement estimates and descriptions.

It is important to implement and maintain effective monitoring and evaluation programs. This follow-up activity is essential to assess the degree to which policy goals are being achieved and to provide a measure of the health status of the environment. It also encourages the selection and measurement of key parameters that provide the desired information to document the degree of risk reduction.

Discussion

Asked whether local people historically are reluctant to accept outside experts, Ron Meekhof responded that many opportunities exist for input from local stakeholders other than agricultural producers. This has provided a positive result with regard to enhanced local participation and acceptance. In response to another question, the speaker indicated that even without financial support from Congress, the USDA is in the process of issuing guidelines on risk assessment/risk management for use at the site level. In Meekhof's view, it would be helpful if the USDA included more quantification to facilitate trade-off analysis through multi-criteria decision making. Discussing the use of radiation to minimize *e-coli* contamination of food, he observed that as the public becomes more aware of the risks remaining in food safety under present inspection programs, support will increase for the USDA to implement more effective protection methods, including radiation of certain foods.

Successes and Failures of Regulatory Risk Analysis in Environmental Standards

William Rowe, President, Rowe Research and Engineering Associates

Bill Rowe reviewed more than twenty-five years of his experience in the application of regulatory risk analysis to environmental standards. He presented information on ten case examples from the enhanced risk of radon in natural gas following underground storage of the gas prior to distribution to developing standards for disposal of radioactive waste. These case examples provided rich insights into the processes of decision making within the US EPA during its formative years as well as the strengths and limitations which the emerging field of environmental risk analysis brings to the policy arena.

Discussion

Commenting on how decision making can be improved, Rowe observed that the US EPA has a difficult task in the open hearing process. Political aspects associated with issues raise the question of whether or not one can perform an objective risk analysis. This analysis becomes a value/choice exercise as soon as information is invalidated. The answer is to state that this is the best information presently available, and proceed to show the implications of various alternatives.

Is a good job currently being done in communicating risk concepts to decision makers and politicians? Rowe responded to this question with a clear "NO." It is imperative that the risk analysis process communicate to them the consequences of choice.

On the issue of dam safety and how it should be addressed, Rowe observed that the parties responsible for dam safety want to avoid catastrophic failures which may occur as a consequence of either extreme natural events or of human error. The risk analysis needs to focus on steps to mitigate the consequences of natural disasters, as well as on what should be done to avoid human error.

Regarding the balance between economic cost vs. social judgment, Rowe believes that our system needs to be modified so that we do not focus so exclusively upon very-rare-event issues. Asked whether or not it is anticipated that regulatory reform will have a positive impact upon risk analysis, he observed that in a technical sense, it should. However, it is important to recognize that for many years the concept of risk has been hidden behind the concept of standards. Accordingly, in the past, the regulators and the public took comfort in the fact that if the standards were being met, the risk would be low.

Is Regulatory Planning and Decision Making an Oxymoron?

Eugene Stakhiv, US Army Corps of Engineers

Gene Stakhiv's presentation challenged the Conference to re-examine the concept of integrated multi-objective analysis, which provided for the systematic examination of the trade-off between economic goals, environmental goals, and human needs. He focused upon risk analysis evolving into a decision-making protocol whereby the resulting regulatory model dominates water-resource planning and development. Stakhiv sees the application of risk analysis techniques contributing more constraints (more regulations), rather than encouraging active trade-offs between multiple objectives. Protection, development, and management activities in the context of multiple-objective river basin development each have their own set of regulatory constraints. His triple-axis representation showed the following principle components: planning vs. regulation, reliability analysis vs. risk analysis, and multi-objective trade-off vs. no trade-off. The region defined by regulation, reliability analysis, and no trade-off is shown to include safety, dam safety, health, nuclear power, drinking water, and dredge material. Stakhiv went on to develop the point that risk analysis is one tool for evaluating options, but not the only effective technique for project evaluation.

The speaker concluded with a discussion of the NEPA process. He identified certain concerns that presently appear to bias the results of the NEPA analysis to favor environmental quality over national economic development.

Discussion

Session participants agreed on the importance of multi-objective analysis to provide full consideration to the trade-off between objectives. It was suggested that risk analysis needs to be tailored to fit different locations within the multi-objective analysis. Risk analysis is another tool to be used in alternative assessment and evaluation. It was also shared during the discussion period that the President's Commission on Risk Management had made similar recommendations. Finally, it was observed that in at least some cases, environmental regulations evolved into constraints because the responsible agency failed to apply appropriate multi-objective analysis and management.

SUMMARY OF RESPONSES
TO PARTICIPANT QUESTIONNAIRE

A questionnaire was distributed to all conference participants, requesting answers to each of the following questions. A listing of the participants' responses follows each question.

l. List the three most important issues/aspects/elements related to risk-based decision making that were raised during this conference.

- The use of expert opinion
- Should we have definite standards in risk or not?
- Dam-safety issues
- Safety levels for hydraulic structures
- Some participants were uninformed about past efforts; some have retrogressed
- New awareness in risk and uncertainty
- Risk analysis is becoming overly institutionalized in federal agencies, especially EPA!
- How to bring risk assessment into the regulatory process
- Growth of acceptance of risk analysis
- Lack of government support for risk analysis
- Comparative risk analysis was not discussed
- Methods are developed between federal agencies without reading the literature
- We have to learn more from each other
- Necessity to manage uncertainty – consequences
- Quantitative modeling in ecosystem protection and quality for risk analysis/risk management
- Willful and purposeful hazards to water resources as a risk assessment/risk management issue
- TRIZ
- Lack of multiobjective decision making
- Roles of uncertainty, variability, and bias in decision making
- Safety issues
- PDF specification
- Extreme-event analysis

- Time-dependent modeling needs attention
- Models for evalution are being developed
- What is the value of judgment in risk-based decision making?
- How do decision makers value risk-based decision making?
- How does this information influence different kinds of decisions?
- Credibility does not come from science or rigor, so where does credibility enter?
- Communication of results needs work
- Monte Carlo simulation
- Time-dependent reliability and hazard function
- Economic cost-benefit analysis
- Risk aversion for extreme events
- Fuzzy-rule-based modeling
- Should risk assessment be primary task?
- The need of broader considerations
- Notion of how safe is safe!!
- Risk and trade-offs
- Decision criteria
- More explicit practical methods to express uncertainty vs. variability
- Misunderstanding of ecologic risk assessment in real world
- The attempts of multiple agencies to define acceptable risk as a benchmark
- Multiobjective analysis/selling of risk analysis to the public
- Comparing risks in daily life with extreme events and seeing how small they really are
- Noticeable lack of improvement in risk analysis in the last decade
- General applicability of fuzzy logic and sets to risk-based decision making
- TRIZ
- Links to environmental risk assessment
- New methods in handling uncertainty
- Fuzzy sets as applied to water resource problem
- Extreme-event analysis
- We are falling behind!! Retrogressing simple mindedly...

2. List the three most important new ideas/concepts that you have learned during this conference that would be helpful in your job.

- Extreme-event modeling
- Fuzzy-rule-based modeling
- Uncertainty in health risk analysis
- Cost-benefit analysis
- Level of uncertainty really impacts on the level of analysis that is conducted

- Expert judgment is much more difficult to handle than one recently thought
- Maximum entropy/probability bounds approach
- Regulatory approaches
- Need to characterize uncertainty
- Use of multivariate analysis in risk prioritization
- Strategies for managing uncertainty
- The rehabilitation of structures is a gigantic multiobjective dynamic problem under uncertainty
- The methodological field must be enlarged not only to statistical and Bayesian methods, but to fuzzy sets, rough sets, belief functions, and possibility theory, among others
- A few tutorials may be in order – but not at too low a level, please!
- TRIZ
- Rowe's hierarchy of uncertainty – good taxonomy
- Classifying robust techniques for uncertainty measurement
- Interaction of connecting water barrier and levees in Netherlands
- Systematic planning of navigation, locks and dams, dredging
- Multiobjective analysis
- Extreme-value analysis
- Keep the "experts" away from your "decision makers"
- Degree of belief is a key element in risk assessments
- Need to focus on gathering useful information to support decisions
- Status of dam safety standards in US and Canada
- The Dutch approach to issue of sea defense
- Update on perspectives of planning/evaluation/risk for water resources at the US federal level
- Criteria-based ranking for environmental investments
- Extreme precipitation is related to both ENSO and circulation patterns
- "Uncertainty" vs. "risk" vs. "surety"
- Dispersive Monte Carlo sampling
- Wicked decisions
- Competent error vs. negligent error
- How to deal with risks and uncertainties with case studies
- Dealing with expert opinion and its role in risk analysis
- Similarity of risk analysis issues across different sources (areas) of decision making
- QA for Monte Carlo
- Risk and uncertainty management strategies
- Definition of terms
- Uncertainty relative to risk
- Risk assessment vs. risk management
- Methods for acceptable risk determination

- "Probability bounds" methods may be helpful to generate input distributions when you don't know much
- Use of evacuation models
- Risk management of portfolios
- New means for combining distributions
- Range estimates of parameters and results are required rather than point estimates
- New approaches to dealing with uncertainty
- Use of fuzzy sets in climate and weather research

3. List the three most important issues needing further study in risk-based decision making in water resources.

- Fuzzy modeling
- Handling the public perception of risk
- How analysis is acquired and used in practice
- Presenting uncertainty to the decision makers
- Bayesian methods in cost-benefit analysis
- Selecting distributions for extreme events
- Defining decision making
- How to be right rather than correct
- How to better choose/revise measurement endpoints for risk analysis/risk management
- Modeling purposeful attacks on water systems
- The role of risk analysis in a total system analysis
- How government/industry are setting acceptable risk levels
- Relevance of Bayesian theory to decision makers
- Uncertainty analyses
- Dealing with long-term, cumulative, uncertain risks
- Regulatory reform
- How to fund risk-based decision making in public agencies
- Regulatory reform
- Role of risk analysis as part of multiobjective decision making – not vice-versa
- Education/training for field analysts: top 10 books/software needed; what are colleges/industry/government doing?
- Extreme-event analysis
- Fuzzy-set workshop (go through application)
- Dam safety/sea level rise
- How to convey results of risk-based decision making analyses to the decision maker
- Agreement on terminology
- Dam safety guidelines
- Multiple failure modes
- Regulating problems

- Society's willingness to pay to mitigate risks
- Comparison of subjective probability vs. quantitative
- Risk and cost of mis-specified models (wrong model)
- Need to establish benchmarks vs. aversion to benchmarks
- General agreement among the members on the framework for risk analysis
- To understand how each organization's standard compares to the benchmark
- To consider risk as just one element of decision making
- Providing the use of good science in practice
- Encoding uncertainties in risk analysis
- How are trade-offs made reliably?
- Use of good systems engineering to define risk problems, and then assess and manage them
- Bridge the gap between "soft" and "hard" versions of risk analysis (polity and social issues vs. engineering/math models) – significant, possible, probable
- Provide predictive studies with validation and honest constructive criticism of past mistakes
- Evaluation and perception of risk compared to other risk sources
- Bayesian or fuzzy techniques
- Make uncertainties in risk assessment visible
- Get quantitative results instead of conceptual/general models
- Ways to characterize risks due to human operator error
- Current state-of the-art paper on expert elicitation
- Fuzzy sets
- Value of standard basis analysis
- Bias
- Risk-based water management strategies and practices
- Incorporate multiobjective analysis into risk analysis
- Ways of presenting uncertain information to decision makers and the public
- Means to make past history and development of terms, methods, and strategies available to more people (especially new ones!) in the field
- Publish more copies of proceedings so they don't go out of print so fast!
- How to arrive at acceptable level for risk management
- Rule curves
- Use multiobjective framework
- Reflecting uncertainty in forecasts of demands
- System model that incorporates the multiobjective, multivariate nature of water resource issues
- To what extent is uncertainty information desired by decision makers? Do they really want it, or are we giving it to them because *we* think they should have it?

PARTICIPANTS

ENGINEERING FOUNDATION CONFERENCE
RISK-BASED DECISION MAKING IN WATER RESOURCES VIII

October 12-October 17, 1997
Santa Barbara, California

Anderson, Elizabeth
  Sciences International, Inc.

Andreasen, James K.
  US Environmental Protection Agency

Bogardi, Istvan
  University of Nebraska

Bulkley, Jonathan W.
  University of Michigan

Cattanach, Dave
  BC Hydro

Chartier, Antoinette
  Engineering Foundation

Daniel, Robert
  US Army Corps of Engineers

Den Heijer, Frank
  Delft Hydraulics

Dise, Karl M.
  US Bureau of Reclamation

Duckstein, Lucien
  University of Arizona

Ezell, Barry C.
    University of Virginia

Ferson, Scott
    Applied Biomathematics

Foster, Jerry L.
    US Army Corps of Engineers

Glover, Terry
    Utah State University

Haimes, Yacov Y.
    University of Virginia

Hennig, Charles
    Department of the Interior

Kaplan, Stanley
    Bayesian Systems, Inc.

Kiefer, Jack C.
    Planning and Management Consultants

Krouse, Michael
    US Army Corps of Engineers

Lambert, James H.
    University of Virginia

Leggett, Mary Ann
    US Army Corps of Engineers

Meekhof, Ronald L.
    US Department of Agriculture

Moser, David A.
    US Army Corps of Engineers

Muller, Bruce
    US Bureau of Reclamation

Nanda, S.K.
    US Army Corps of Engineers

O'Connor, Robert E.
Pennsylvania State University

Patev, Robert C.
US Army Engineers Waterways Experiment Station

Pikus, Irwin M.
President's Commission on Critical Infrastructure Protection

Powell, Mark R.
US Department of Agriculture

Redding, John
City of Thornton, Colorado

Roos, Alex
Ministry of Transport, Public Works and Water Management, Delft

Rowe, William
Unlimited Assurance Holdings, Inc.

Smart, John D.
US Bureau of Reclamation

Stakhiv, Eugene Z.
US Army Corps of Engineers

Stott, Charles B.
Engineering Foundation

Van Gelder, Pieter
Delft University of Technology

Walker, Wesley
US Army Corps of Engineers

Watson, Robert L.
University of Virginia

Wolfson, Lara J.
University of Waterloo, Canada

Yoe, Charles E.
College of Notre Dame of Maryland

ENGINEERING FOUNDATION CONFERENCE

on

RISK-BASED DECISION MAKING IN WATER RESOURCES VIII

Radisson Santa Barbara Hotel
Santa Barbara, California

October 12-17, 1997

Co-sponsor:
The Universities Council on Water Resources (UCOWR)

Supported by:
National Science Foundation
US Army Corps of Engineers

Conference Chairman: Yacov Y. Haimes, University of Virginia
Conference Co-Chairman: David Moser, US Army Corps of Engineers,
Institute for Water Resources

Proceedings Editors:
Yacov Y. Haimes, David A. Moser, and Eugene Z. Stakhiv
Technical Editor:
Grace I. Zisk

DECISION MAKING IN WATER RESOURCES VIII

SUNDAY, October 12, 1997

3:00 p.m. - 9:00 p.m.   REGISTRATION

6:00 p.m.   DINNER

7:30 p.m. - 10:00 p.m.   SOCIAL HOUR

MONDAY, October 13, 1997

7:00 a.m. - 8:00 a.m.   BREAKFAST BUFFET

8:30 a.m. - 12:00 noon   **SESSION I: PROTECTION OF CRITICAL INFRASTRUCTURES**
Plenary Session Chair: Yacov Y. Haimes, University of Virginia

PRESENTATION OF PAPERS

Overview and Findings: President's Commission
Irwin Pikus, Commissioner, President's Commission on Critical Infrastructure Protection

Hardening of Water Supply Systems
James Lambert, University of Virginia

Vulnerability of SCADA Systems
Barry Ezell, University of Virginia

Risk of Power and Water Interfaces
Robert Watson, University of Virginia

10:30 a.m. - 11:00 a.m.   COFFEE BREAK

12:00 noon - 1:00 p.m.   LUNCH

1:00 p.m. - 5:00 p.m.   Ad Hoc Sessions and/or Free Time

5:00 p.m. - 6:00 p.m.   SOCIAL HOUR

6:00 p.m. - 7:00 p.m.   DINNER

7:00 p.m. - 9:30 p.m.

SESSION II: DAM SAFETY AND RISK OF EXTREME AND CATASTROPHIC EVENTS
<u>Session Chair</u>: David Moser, US Army Corps of Engineers
<u>Rapporteur</u>: Wesley Walker, US Army Corps of Engineers

PRESENTATION OF PAPERS

Risk-Based Decision Making for Dam Safety
Jerry Foster, US Army Corps of Engineers

Dam Safety Evaluation and Decision Making in Australia
Terry Glover, Utah State University

Risk-Based Dam Safety Evaluations and Improvements
Dave Cattanach, BC Hydro

Achieving Public Protection in Dam Safety Decision Making
Chuck Hennig, US Bureau of Reclamation

TUESDAY, October 14, 1997

7:00 a.m. - 8:00 a.m.

BREAKFAST BUFFET

8:30 a.m. - 12:00 noon

SESSION III: NAVIGATION SYSTEMS
<u>Session Chair</u>: S.K. Nanda, US Army Corps of Engineers
<u>Rapporteur</u>: Barry Ezell, University of Virginia

PRESENTATION OF PAPERS

**Towards a Risk-Based Analysis –
Ohio River Navigation Study**
Wesley Walker, US Army Corps of Engineers

Emerging Corps Policy – Navigation
Robert Daniel, US Army Corps of Engineers

A Systems Approach to the Optimal Safety Level of Connecting Water Barriers in a Sea-Lake Environment; Case Study: Afsluitdijk, The Netherlands
Alex Roos, Ministry of Transport, Public Works and Water Management, Delft, The Netherlands

Risk-Averse Reliability-Based Optimization of Sea Defenses
Pieter van Gelder, Delft University of Technology

10:30 a.m. - 11:00 a.m.	COFFEE BREAK
12:00 noon - 1:00 p.m.	LUNCH
1:00 p.m. - 5:00 p.m.	Ad Hoc Sessions and/or Free Time
5:00 p.m. - 6:00 p.m.	SOCIAL HOUR
6:00 p.m. - 7:00 p.m.	DINNER

7:00 p.m. - 9:30 p.m.   **SESSION IV: NSF SECOND WORKSHOP ON "WHEN AND HOW CAN YOU SPECIFY A PROBABILITY DISTRIBUTION WHEN YOU DON'T KNOW MUCH?"**
Session Chair: James Lambert, University of Virginia
Rapporteur: William Rowe, Unlimited Assurance Holdings, Inc.

The Role of Uncertainty, Variability, and Bias in Risk Management
Elizabeth Anderson, Sciences International, Inc.

You Don't Need to Specify Precise Distributions
Scott Ferson, Applied Biomathematics

On the Application to Risk and Decision Analysis of TRIZ, the Russian Theory of Inventive Problem-Solving
Stan Kaplan, Bayesian Systems, Inc.

Methods for Characterizing Variability and Uncertainty: Bayesian Approaches and Insights
Lara Wolfson, University of Waterloo, Canada

WEDNESDAY, October 15, 1997

7:00 a.m. - 8:00 a.m.   BREAKFAST BUFFET

8:30 a.m. - 12:00 noon   **SESSION V: FORECASTING IN MANAGEMENT OF WATER RESOURCES**
Session Chair: Istvan Bogardi, University of Nebraska
Rapporteur: Charles Yoe, College of Notre Dame of Maryland

PRESENTATION OF PAPERS

A Fuzzy Rule-Based Model to Link Circulation Patterns, ENSO, and Extreme Precipitation: An Arizona Case Study
Lucien Duckstein, University of Arizona

Current Corps Research Efforts on Reliability and Risk Analysis
Mary Ann Leggett, US Army Corps of Engineers

Strategies for Managing Uncertainty
William Rowe, Unlimited Assurance Holdings, Inc.

Risk of Regional Drought Influenced by ENSO
Istvan Bogardi, University of Nebraska

10:30 a.m. - 11:00 a.m.   COFFEE BREAK

12:00 noon - 1:00 p.m.   LUNCH

1:00 p.m. - 5:00 p.m.   Ad Hoc Sessions and/or Free Time

5:00 p.m. - 6:00 p.m.   SOCIAL HOUR

6:00 p.m. - 7:00 p.m.   DINNER

7:00 p.m. - 9:30 p.m.   **SESSION VI: ECOSYSTEM PROTECTION**
Session Chair: Eugene Stakhiv, US Army Corps of Engineers
Rapporteur: Robert O'Connor, Pennsylvania State University

PRESENTATION OF PAPERS

**Criteria-Based Rankings in Risk-Based
Analysis of Environmental Investments**
Charles Yoe, College of Notre Dame of Maryland

**EPA's Proposed Guidelines for Ecological Risk
Assessment**
James Andreasen, US Environmental Protection
Agency

Economic Issues in Environmental Regulations
Mark Powell, US Department of Agriculture

**Extreme Value Analysis of Waste Water
Treatment Plants**
Jonathan Bulkley, University of Michigan

THURSDAY, October 16, 1997

7:00 a.m. - 8:00 a.m.   BREAKFAST BUFFET

8:30 a.m. - 12:00 noon   **SESSION VII: REGULATORY ANALYSIS**
                               Session Chair: Michael Krouse, US Army Corps of
                               Engineers
                               Rapporteur: Jonathan Bulkely, University of Michigan

                               PANEL DISCUSSION

                               **Risk Assessment of USDA and Conservation
                               Programs**
                               Ronald Meekhof, United States Department of
                               Agriculture
                               S.K. Nanda, US Army Corps of Engineers
                               Eugene Stakhiv, US Army Corps of Engineers

10:30 a.m. - 11:00 a.m.   COFFEE BREAK

12:00 noon - 1:00 p.m.   LUNCH

1:00 p.m. - 5:00 p.m.   Ad Hoc Sessions and/or Free Time

5:00 p.m. - 6:00 p.m.   SOCIAL HOUR
6:00 p.m. - 7:00 p.m.   DINNER

7:00 p.m. - 9:30 p.m.   **SESSION VIII: ADVANCES IN RISK ANALYSIS**
Session Chair: Mary Ann Leggett, US Army Corps of
Engineers
Rapporteur: Charles Stott, Engineering Foundation

PRESENTATION OF PAPERS

**Putting Water Risks in Context: Public
Perceptions and Their Implications for Risk-Based
Decision Making**
Robert O'Connor, Pennsylvania State University

Extreme Events
Yacov Y. Haimes, University of Virginia

**Time-Dependent Reliability Analysis and Hazard
Function Development for Civil Works Structures**
Robert C. Patev, US Army Engineers Waterways
Experiment Station

FRIDAY, October 17, 1997

7:00 a.m. - 8:00 a.m.   BREAKFAST BUFFET

8:30 a.m. - 10:00 a.m.   Summary of Questionnaire
Yacov Y. Haimes, University of Virginia

10:00 a.m. - 10:30 a.m.   CONCLUSION AND ADJOURNMENT

Subject Index

Page number refers to the first page of paper

Author Index

Page number refers to the first page of paper